PROCESSES OF ANIMAL MEMORY

PROCESSES OF ANIMAL MEMORY

Edited by

Douglas L. Medin
THE ROCKEFELLER UNIVERSITY

William A. Roberts
UNIVERSITY OF WESTERN ONTARIO

Roger T. Davis
WASHINGTON STATE UNIVERSITY

LAWRENCE ERLBAUM ASSOCIATES, PUBLISHERS
1976 Hillsdale, New Jersey

DISTRIBUTED BY THE HALSTED PRESS DIVISION OF
JOHN WILEY & SONS
New York Toronto London Sydney

Lawrence Erlbaum Associates, Inc., Publishers
62 Maria Drive
Hillsdale, New Jersey 07642

Distributed solely by Halsted Press Division
John Wiley & Sons, Inc., New York

Library of Congress Cataloging in Publication Data
Main entry under title:

Processes of animal memory.

Expanded versions of papers presented at a conference held Oct. 16–18, 1974, at Augustana College, Rock Island, Ill.

Bibliography: p.
Includes indexes.

1. Memory–Congresses. 2. Learning, Psychology of–Congresses. 3. Animal intelligence–Congresses.
I. Medin, Douglas L. II. Roberts, William A.
III. Davis, Roger T.
QL785.P76 156'.3'12 76-22197
ISBN 0-470-15189-7

Printed in the United States of America

Contents

Preface

In October, 1974, a conference on animal memory was held at Augustana College in Rock Island, Illinois. The papers presented in this volume represent expanded versions of the presentations made at that conference, along with two contributions that were invited subsequent to the conference. Since no constraints were placed upon the topics of these papers, other than that they be concerned with animal memory in some way, the resulting collection represents a diverse but representative sample of current research on memory in animals. Despite their variety, a reading of these chapters suggests that many ideas, problems, and theoretical points of view are common to them. A delineation of some of these currents of thought may help us to understand where the present interest in animal memory has come from and where this field may be heading.

First, we see an interest in the comparative aspects of memory. This interest is virtually inherent in animal research and in the present volume seems to take two principal forms. In the contributions by Roberts and Grant and of Medin, an approach to comparative analysis is suggested in which theoretical models of memory are constructed for different species from empirical data and then are compared across species. Such a venture is held to be useful in discovering possible differences in memory processes between different species of animals (and between animals and man), between ontogenetically distinct groups (Spear and Parsons' chapter), and in refining theories of memory within any single species. The other comparative aspect of these papers reflects the current emphasis upon constraints in learning. Bolles suggests that animals may be genetically limited in the kinds of memories they can form and retrieve, and D'Amato and Cox suggest that monkeys may not have evolved the processes necessary for foresightful behavior.

It is undoubtedly true that the enormous expansion of research on human memory over the last 15 years or so has rubbed off on the researcher concerned with animal learning. Thus, we find problems and paradigms used in human memory being translated into animal experiments. A good example of this is the recent interest in short-term memory in animals, which is reflected in the papers of Roberts and Grant, Ruggiero and Flagg, Medin, and Davis and Fitts. Interest in the effects of spaced repetition upon retention, retention of a "list" of stimuli, and proactive inhibition effects in short-term memory all represent extensions of problems first studied with human subjects.

A closely related trend is an eagerness to examine the relationship between memory processes and other cognitive processes in animals. This has led to a realization that many of the traditional problems in animal learning might be better understood if theoretical analyses at least partially based upon memory processes were brought to bear upon them. Thus Bolles argues that just as a learning analysis of memory paradigms proved of value in the past so also may a memory analysis of learning paradigms now prove fruitful. Of particular interest in the Bolles paper is his discussion of the Garcia effect (long delay taste aversion learning) and his suggestion that selective retrieval mechanisms and specialized memory buffers may have evolved to suit particular needs.

The papers of Whitlow, Davis and Fitts, Bartus and LeVere, and D'Amato and Cox reveal also the use of memory mechanisms to deal with old problems in animal learning. In a review of work done recently at the Yale laboratories, Whitlow shows how a postperceptual processing or rehearsal mechanism in animals can explain a number of findings in classical conditioning. In several ways, the paper of Bartus and LeVere represents a similar analysis of memory or information processing in instrumental discrimination learning. Both the Whitlow and Bartus and LeVere papers argue that learning is not completed when the events of a trial are finished but involves the continuation of information processing which will determine what and how much is learned. The paper by D'Amato and Cox attacks the traditional problem of delay of reward; based upon the monkey's good retention over periods of minutes in delayed matching to sample, they ask if the monkey cannot learn to delay reward for similar lengths of time and still respond accurately on visual discriminations. The reader of this paper will realize quickly that there is a substantial difference between asking a monkey to delay response and asking it to delay reward. Davis and Fitts show that the answers to questions about the relative effectiveness of reward and nonreward depend critically on the retention interval. They explore the implications of these findings for Harlow's uniprocess theory of learning set formation.

Finally, all of the chapters reflect an interest in memory *processes.* As Bolles points out, taste aversion learning over long delays is not significant for the delays per se, but is important because of what learning over such delays implies about the information processing activities of the organism. Ruggiero and Flagg distinguish between S–R, representational, and organized memory in animals, the latter two corresponding approximately to Tulving's distinction between episodic and semantic memory. Ruggiero and Flagg's analysis of types of memory serves to direct our attention away from concern with paradigms in favor of a focus on processes. We think the chapters in this book make an important contribution to elucidating processes of animal memory.

ACKNOWLEDGMENTS

This book is an outgrowth of a conference on animal memory that was held at Augustana College in Rock Island, Illinois, October 16–18, 1974. The College supplied a grant to Frank T. Ruggiero in the psychology department for lodging, meals, and a place for the conference to meet in the gracious Weyerhouser mansion, and the Parke-Davis Drug Company on a request from Raymond T. Bartus bore the costs of air transportation of the participants to Rock Island.

Among our hosts at Augustana were President C. W. Sorensen and Dr. Ross Paulson, Chairman of the Division of Social Sciences. Dr. Duncan McCarthy, Director of Pharmachology for the Parke-Davis Research Laboratories, saw the importance of studying memory in animals for ultimately dealing with memory disorders in the retarded and aged. It was through his interest and the need for the expression of new ideas that we received support from Parke-Davis. We therefore dedicate this book to our friends at Augustana and the Parke-Davis Company.

Finally, the editors would be remiss if they did not acknowledge the great pleasure of working with a very patient, understanding, gentleman and scholar, Lawrence Erlbaum, the publisher.

DOUGLAS L. MEDIN
WILLIAM A. ROGERS
ROGER T. DAVIS

PROCESSES OF ANIMAL MEMORY

1

Do Animals Have Memory?

Frank T. Ruggiero[1]
Steven F. Flagg[2]

Washington State University

INTRODUCTION

Common experience indicates that animals remember things—dogs know their masters, homing pigeons can find their roosting place, bait-shy rats avoid poisoned food. Therefore, the following anecdote may come as a surprise. Several years ago a fellow graduate student was engaged in defending his thesis, which was concerned with imprinting in chickens. At the outset, the student discussed changes in the animals' memory that had occurred during the experiment. One of the student's committee members was from the philosophy department and midway through this presentation the philosopher stopped the discussion, proclaiming that he could not understand why the student persisted in using the word memory to describe the animals' behavior. After all, he pointed out, "Aristotle has told us that animals are incapable of higher processes and therefore, they cannot possess memory!" Having said that, he left.

To better understand the philosopher's consternation and the purpose of this chapter it is necessary to consider briefly the historical context of animal memory. Since the time of Aristotle, man has been regarded as different from animals because he possesses a reasoning ability, based on a spoken language. Thinking was, and often still is, equated with linguistic ability. For instance, until the 16th century, when a Spanish nobleman's deaf son was taught to read and write, the deaf were legally considered to be subhuman because they did not manifest speech (Furth, 1966). Such thinking reflects the view that, man, by virtue of his verbal ability, is capable of actively recalling the past (i.e., generating information about a past event in the absence of the original event), whereas nonverbal organisms are capable of recognition at best. If the term "memory" is

[1] Currently at Augustana College, Rock Island, Illinois, 61201.

[2] Currently at the University of Maryland, College Park, Md. This chapter was written while the second author was at Hope College, Holland, Mich.

restricted to instances of active recall, we can begin to understand the philosophy professor's claim that animals do not have memory.

In this chapter we examine three distinct types of memory, the first two corresponding roughly to recognition and recall. We do this not because we hope to deny or delimit mnemonic capabilities in animals, but because different theoretical questions arise in considering these different kinds or aspects of memory.

STIMULUS–RESPONSE MEMORY

The first type of memory we consider, S–R memory, has often gone under the designator "retention." Because learning would be impossible without some memory, the study of changes in the strength of associations comprising learning was compatible with the S–R learning theories of Hull (1943) and Spence (1936, 1937) which dominated research in the 1930s and 1940s. The S–R bond was all important, and because memory was not defined independently from the S–R bond the use of the term "memory" seemed to be an unnecessary additional hypothetical construct. Learning experiments examined the formation of the S–R bond and retention tests, when given, were concerned with determining how long this S–R bond lasted. (See chapter 2 for further analysis of the influence of the learning orientation on the study of memory.) For the purpose of this chapter we call this conceptualization of memory *S–R memory.*

S–R memory was investigated in animals using various learning paradigms. Typically an animal was trained to a criterion on a task and then tested or retrained on that task days, months, or years later. If the subject performed better on the latter exposure to the task than control animals with no prior experience on the tasks, the animal was said to have retained the original training. It was soon found that animals showed remarkable retention for well-learned tasks and the investigator's interest often faded at that point. Instances of forgetting were interpreted mainly in terms of interference caused by new and prior learning (see Gleitman, 1971).

In terms of the distinction between recall and recognition, virtually all studies in the S–R framework fall under the domain of recognition—the appropriate stimulus conditions are presented for the retention tests and the animal either does or does not then behave appropriately. In these tasks there is an easily identifiable stimulus that can elicit the response of the animal.

The very nature of the tasks used to study animal memory resulted in animal memory being easily interpreted in the classical S–R terms. No widely used animal paradigm readily comes to mind that corresponds to studies of recall in human subjects. Yet, as we shall see, memory in animals is not limited to the retention of learned acts.

REPRESENTATIONAL MEMORY

Even in the face of S–R associationisms' heavy influence, there was research aimed at determining whether animal memory might consist of more than response tendencies. That is, perhaps animals have an independent representation of a stimulus event, one that is stored in memory and can be used (recalled) in the absence of that stimulus. Memory that is defined independently of the S–R bond and that can be assessed in a number of ways, not just as a conditioned response to a stimulus, shall be referred to as *representational memory*. Historically, Tolman's S–S psychology seems most compatible with representational memory. Indeed, Tolman was criticized for not having ready rules for translating memory into performance.

A clear demonstration of representational memory in animals is difficult because most nonverbal memory tests involve recognition and, by definition, there is always a stimulus present to elicit the response in these tasks. In order to demonstrate representational memory, one must devise a test where the stimulus is not present at the time of the retention test, or one in which there has been no opportunity to form a S–R bond.

Delayed Response Problems

Hunter's (1913) classic work on the delayed-response problem was perhaps the first attempt to determine experimentally whether animals have representational memory. The paradigm was designed to analyze "mammalian behavior under conditions where the determining stimulus is absent at the time of response" (Hunter, 1913, p. 1). To assess whether or not animals could react on the basis of "mental imagery," rats, dogs, raccoons, and children from two age groups were taught to associate a light at one of three food boxes with reward. After this training the light at the food box was extinguished before the subject responded, requiring the subject to remember where the light had been. No response was made at the time of stimulus presentation.

Different time delays were interposed between the offset of the stimulus light and raising the clear glass restraining chamber. Raising the restraining chamber permitted the subject to run to one of the food boxes. Hunter was primarily interested in two things, the maximum delay the different species could wait and how they were mediating the delay. Because the subject had to respond in the absence of the stimulus that had previously guided his reaction (the light), this seemed to Hunter to offer the potential of testing for representational memory in animals. In his words: "If a selective response has been initiated and controlled by a certain stimulus, and if the response can still be made successfully in the absence of that stimulus, then the subject must be using something that functions for the stimulus in initiating and guiding the correct response" (p. 2).

Restated in the terminology used in this chapter, if the stimulus (light) is not there at the time of the response, memory must be more than a S–R bond. There must be a representation of the light held in memory that can be used even when the light is not present.

On the basis of behavioral observations made during the delay period, Hunter concluded that four levels of learning were needed to encompass the learning abilities of animals and children: (1) inability to profit from experience; (2) trial and error; (3) sensory thought; and (4) imaginal representation. He classified the behavior of his rats and dogs as trial and error, concluding that they were only capable of mediating the delay through the use of bodily orientation. Raccoons and young children (2–3 years of age) could use orientation cues or not and still respond well above chance. He felt their behavior could be regarded as an example of "sensory thought." Only in the case of the older children (5–7 years of age) would Hunter grant the ability of imaginal representation in memory.

Body Orientation and Memory

Although Hunter was asking an important question about memory, his conclusions, concerning the importance of bodily orientation for memory in animals, were both premature and unfortunate for the development of animal memory research. His results were biased by the specific nature of his training procedures, which encouraged orienting responses (Weiskrantz, 1968). Ample evidence is available indicating that rats can solve various delayed-response problems without the aid of bodily orientation (Maier & Schneirla, 1935; Ladieu, 1944). In particular, investigators have reported that monkeys and apes can readily solve complex delayed-response problems over long delay periods without the aid of orientation cues (Tinklepaugh, 1928, 1932; Yerkes & Yerkes, 1928; Gleitman, Wilson, Herman, & Rescorla, 1963; Weiskrantz, 1968; Miles, 1971; Medin & Davis, 1974). In fact, the use of body orientation seems to be a nonpreferred strategy used by "less capable animals" (Harlow, Uehling, & Maslow, 1932) and one that is not spontaneously adopted (Nissen, Carpenter, & Cowles, 1936).

If delayed-response performance involves more than S–R memory, then this paradigm may provide an important tool for asking questions about representational memory. However, even though covert or overt orientation does not seem to provide the basis for delayed-response performance, these forms of orientation have always lurked as potentially confounding variables. Several years ago we began some work aimed at examining representational memory in monkeys, with orienting responses controlled. The problem is not so simple as it may seem, for we desire a technique for controlling orientation that does not otherwise disrupt performance. For instance, simply lowering the opaque screen of the Wisconsin General Test Apparatus (WGTA) during delay intervals does not suffice, for Motiff, DeKock, and Davis (1969) have demonstrated that this procedure may disrupt performance simply because lowering the screen normally is a signal that the current trial is over. The next several paragraphs describe our efforts to use alternative techniques to control orientation.

Basic memory as pattern reproduction (MPR) task. Because we will make frequent reference to it, the matrix pattern reproduction task illustrated in Fig. 1.1 will be described in some detail. The visual display equipment consisted of a 4 X 4 display of contiguous, translucent plastic panels. Each panel measured 6.4 X 6.4 X .9 cm and was the front wall or door of a 6.4 X 6.4 X 7.6 cm metal box. The door panel was hinged so that the animal could push on it and thereby gain access to a raisin reward lying on an elevated pad inside the box. The pad raised the food reward above the floor of the box so that the animal could not see it through the slight space at the bottom of the door. There were five lamps per cell, which were used to illuminate the plastic door. They were located at the back of each of the 16 boxes. Each of the lamps in a cell was a different color–aircraft red, aircraft blue, aircraft green, aircraft yellow (Woodson & Conover, 1965), or clear. The colored lights were obtained by placing commercially made rubber filters over clear 28 V lamps. The boxes also contained a microswitch that was activated when the door opened, allowing the light to come back on if the animal chose the previously illuminated cell. The entire 4 X 4 matrix display unit was mounted on a revolving base, in a modified WGTA, so that it could be turned 90° away from the animal to allow the experimenter to bait the correct cells of the matrix display. The stimulus conditions, response

FIG. 1.1 Display conditions for the basic MPR task. The nonshaded cell represents illumination of the stimulus cell with white light, and the dotted lines show the positions of the two-way, mirror glass viewing screen during the trial. For the duration of the delay phase the animal cannot see the matrix display. It is shown here for illustration purposes. [After Borkhuis, Davis & Medin, 1971. Copyright (1971) by the American Psychological Association and reproduced by permission.]

delays, and all other experimental conditions were controlled by feeding a punched card into a 80-column, static card reader which operated a solid state logic system.

On a typical trial, the experimenter programmed the equipment to deliver the appropriate experimental conditions. He then baited the correct cell with a raisin, rotated the display 90° into position facing the animals, and raised the opaque screen of the WGTA to initiate the trial. The particular stimulus configuration could be seen by the animal through the viewing screen, which could be raised immediately or left in place for the duration of a variable response delay. The animal was allowed to respond to one of the cells after the viewing screen was raised. If he chose the correct cell he found a raisin in the cell, and the cell light reilluminated, confirming the response. This sequence is diagrammed schematically in Fig. 1.1.

In the only previously reported pattern reproduction task that controlled for orienting responses in monkeys, Riopelle (1959) lowered the opaque screen of his WGTA during the response delay to eliminate his monkeys' view of his horizontal, 1 X 5 matrix display. He reasoned that if his animals were fixating on the correct square during the delay, this procedure should disrupt performance. The manipulation was so disruptive that Riopelle had to stop the experiment after 10 days of testing because of balks and generally uncooperative behavior. The little bit of data he collected showed a drop from 80% correct, with a view of the stimulus display, to 50% without a view of the display. (Chance in this situation was 20%.) Some of the decrement in the animals' performance may be associated with the fact that lowering the screen was also the signal for the end of the trial.

Toward–away problem. In an attempt to test the generality of Riopelle's findings (Ruggiero, 1974) the rhesus monkeys at Washington State University (Davis, 1974) were presented with one-light problems in the MPR paradigm and their view of the 4 X 4 matrix display was eliminated by turning the box away from the animal during the response delay. On toward trials the matrix display remained in place, facing each subject throughout the trial. On away trials the matrix display was turned 90° from the animal as soon as the stimulus light extinguished, and after a delay of 0, 3, 6, or 12 sec the box was turned back into position facing the animal.

The away procedure was very disruptive to MPR performance; it reduced the level of performance on the other delays from 80% correct to 28% correct, better than doubled the number of errors at all delays, and flattened out the negatively accelerated memory curve normally obtained in this paradigm. Because it was well known that old world monkeys could solve delayed-response problems without the use of bodily orientation cues, it was not clear why a view of the display was so important to maintaining performance.

A second study was conducted to determine whether the performance decrement resulted from insufficient time to encode the stimulus position. The same procedure as described above was used, except that on away trials there was a 1-sec delay before the box was turned out of position. Although performance under the new procedure was slightly better at all delays, the differences were not reliable. In addition, a reexamination of the procedures used in these two studies indicated that turning the display may have been associated with the end of trial sequence.

View, no-view performance. In a later study a novel procedure "not related to the end of trial sequence" was used to avoid the procedural problems of the studies previously described. This was accomplished by assessing the performance level of the rhesus monkeys on a MPR task in which the availability of orientation cues was controlled by a two-way, mirror glass screen (see Ruggiero, 1974, for details). A subtle change in the lighting conditions could be used to eliminate the animals' view of the stimulus display. Each animal was presented with all 16 possible one-light problems under both view and no-view response delay conditions at each of five different delays (0, 1, 3, 6, and 12 sec). On a typical view trial one of the cells of the matrix was illuminated for 500 msec; it was then extinguished and a variable delay period followed. The animals were prevented from responding by the mirror glass screen, but they could see the unlit matrix through the two-way screen because the illumination in the experimental chamber was higher than in the animal's testing cage. The viewing screen was raised at the end of the delay period, and the monkey was allowed to respond to one of the cells. If correct, he was rewarded with a raisin and confirming light. No-view trials differed only in that illumination of the experimental chamber behind the two-way screen terminated with the offset of the stimulus light, eliminating the animal's view of the matrix during the delay.

The percentage of correct responses made by the monkeys for view and no-view trials are shown, as a function of increasing response delay, in Figure 1.2. Performance for view trials was better at all delays than that for no-view trials. However, both functions show a negatively accelerated memory curve, and the high 0-sec performance indicates that the animals have been attending to the stimulus display. Apparently the no-view procedure is not especially disruptive and the difference between view and no-view conditions may reflect the possibility for orientation on view trials.

The main dividend of the no-view procedure is that we can begin to ask questions about the nature of the representation in representational memory. For example, are some cell patterns easier to remember than others? The answer is yes, corner cells are easiest to reproduce, followed by edge cells, whereas center cells are the most difficult to reproduce. This replicates a similar finding with sophisticated, old and middle-aged rhesus monkeys (Medin, 1969) and young, relatively naive pigtail monkeys (Ruggiero, 1974), both involving what amounts to a view condition. Errors also tended to cluster around correct cells in a pattern resembling that reported for view trials by Medin and Davis (1974). Without the no-view control procedure, these results might have been inter-

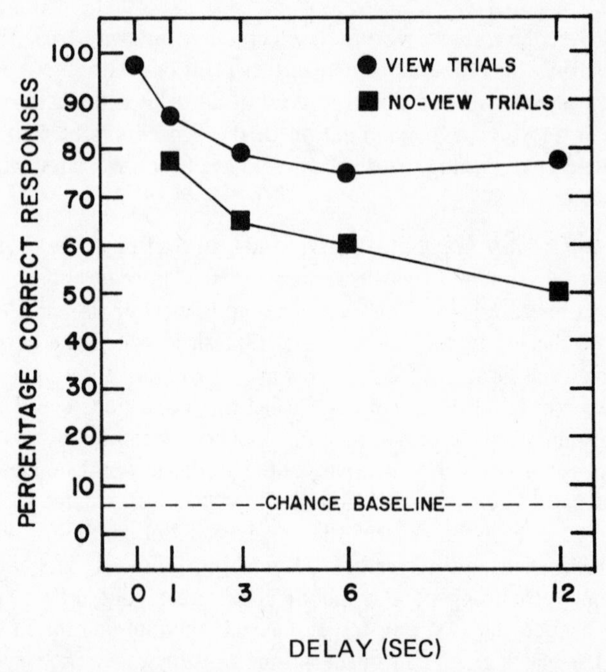

FIG. 1.2 Percentage of correct responses on one-light problems by the rhesus monkeys, as a function of increasing response delay, for view and no-view trials. There is no 0-sec data point for the no-view curve because this condition is not possible to administer.

preted in terms of different possibilities for covert and overt orientation (Fletcher, 1965).

Suppose we try to interpret the no-view study in terms of S–R memory. First, we shall have to deny Hunter's premise that the light is the stimulus and propose that the matrix cells are the effective stimuli. We would assume that in the MPR task the subject builds up 16 S–R bonds or habits, one for each cell. The lighting phase at the start of a trial must be assumed to raise the strength of the correct S–R habit (by its secondary reinforcing properties) above that of the other 15 response tendencies by a large enough amount to produce a correct response at the time of this test.

There are several problems with the notion that the view, no-view experiment involves representational memory rather than S–R memory. For example, the S–R interpretation requires 16 S–R bonds or habits, but in many delayed-response tasks animals perform well above chance before they have had any opportunity to form associations between the stimuli and their unique responses. Direct evidence on this question could be gathered by initially training naive animals on the MPR task using only 12 of the cells. Then memory tests could be run with the other four cells. If the animals do have an independent representation of the stimulus in memory, then they may perform well above chance on those cells, even though they have had no opportunity to form these four S–R associations.

A greater problem with the S–R interpretation of the delayed response task is that if the light temporarily raises one S–R habit, its strength must decline in a negatively accelerated manner and remain high for some time, because monkeys perform well above chance over long delays. In that case, one should find considerable proactive interference from prior trials in the MPR paradigm but to date we have found very little interference (for a review, see Medin & Davis, 1974), even when the interfering stimulus is presented within a single trial as in the unpublished study mentioned in Medin's discussion of interference in Chapter 5 of this volume. We turn now to other types of evidence for representational memory, which also provide information concerning the nature of the representation.

Substitution Tasks

Studies in which nonpreferred rewards are substituted for preferred rewards clearly show that animals have representational memory. The earliest demonstration was conducted by Tinklepaugh (1928, 1932), using the direct delayed-response paradigm in a series of imaginative experiments. In the first of two extensive reports, Tinklepaugh (1928) trained four macaque monkeys to recover a piece of banana that had been placed out of sight under one of two cups. In the basic procedure the animal sat in a chair and watched the experimenter bait one of the cups. For short delays a board separating the animal from the cups was raised, which obstructed his view of the cups during the response delay. After the delay the board was lowered and the animal was allowed to respond by turning over one of the cups. If he was correct, the animal found the food reward under the cup and was allowed to eat it. For long delays the animal was taken out of the room and returned at a later time. Under these conditions the monkey's performance was well above chance even for delays as long as 15–20 hr. However, as was pointed out previously, it might not be necessary to invoke any kind of internal representation in delayed response tasks, so Tinklepaugh modified his basic procedure. During the delay period, he substituted a nonpreferred food for the original preferred food and observed the animal's responses in this situation.

On trials in which the substitution method was utilized Tinklepaugh (1928) reports the following behavior for an animal following a correct response:

> She jumps down from the chair, rushes to the proper container, and picks it up. She extends her hand to seize the food. But her hand drops to the floor without touching it. She looks at the lettuce, but (unless very hungry) does not touch it. She looks around the cup and behind the board. She stands up and looks under and around her. She picks the cup up and examines it thoroughly inside and out. She has on occasions turned toward observers present in the room and shrieked at them in apparent anger. After

several seconds spent searching she gives a glance toward the other cup, which she has been taught not to look into, and then walks off to a nearby window. To prove that lettuce is still appetizing to her, a simple delay is tried in which the same lettuce she had just rejected is used for reward. She responds normally, takes the lettuce, and eats it [pp. 224–225].

This is clearly the behavior of an animal that expected a preferred food and was surprised when this food was not there, not an animal simply running off a S–R habit. A similar reaction was also obtained when the amount of food was changed during the delay. Tinklepaugh concluded that this type of behavior indicated that monkeys were capable of higher representational processes (i.e., the subject had a representation of the reward in memory that could be retrieved and used even in the absence of that reward). Tinklepaugh (1932) repeated this work with chimpanzees and found the same results, although the chimpanzees performed at a higher level than his macaque subjects. If the reader is critical of Tinklepaugh's interpretation, he is reminded that recently Bower (1971) and his associates have used a similar substitution method and the same logic to great advantage in examining the nature of object permanence and memory in human infants.

A study by Medin (cited in Medin & Davis, 1974) provides a more objective measure of reward expectancy than the emotionality measure of Tinklepaugh. Medin reasoned that if monkeys had some kind of representation of a reward in memory, then response speeds might vary as a function of reward preference. The actual task simply consisted of having rhesus monkeys displace a single object in order to retrieve a reward consisting of a preferred food (raisin) or a nonpreferred food (celery). At the beginning of each trial the foodwell area was flooded for two seconds with either red light or green light, depending on the nature of the reward under the object. Five seconds after the colored cue light was extinguished a clear screen that had prevented the monkey from responding was removed. At the time of the response, therefore, the stimulus situation was exactly the same regardless of whether the preferred or nonpreferred food was under the object.

The results were clearcut and suggest that the monkeys were using representational memory. The animals responded reliably faster when the cueing light indicated that a preferred food was available, than when the light indicated that a nonpreferred food was available. (It should be noted that no animal ever failed to eat the nonpreferred celery reward.) In short, the monkeys' response speeds were directly related to the quality of reward that a color cue signaled was available. Because the stimulus situation at the time the monkeys were allowed to respond was identical regardless of which reward was forthcoming, it is impossible to explain the difference in response speeds via S–R memory. There simply were no differential stimuli at the time of the response to elicit different responses. The monkeys had to hold some kind of representation concerning the nature of the reward in memory and then be able to base their response on that representation.

Alternation Problems

In alternation problems subjects must alternate their responding between one location and another location (spatial alternation), or between one object and another object (nonspatial alternation). Double-alternation problems are similar except that subjects must respond to one stimulus (or location) twice, then to the other stimulus (or location) twice, and then to the first stimulus twice, and so on. If delays are interposed between responses the paradigm becomes a delayed-alternation problem. Monkeys are quite capable of solving all of these problems (see French, 1965, for a thorough analysis of this type of problem). However, as French correctly indicates, "The fact is that the sources of cues to successful delayed alternation are today neither intuitively obvious nor experimentally demonstrated [p. 181]." The reason for this statement is that the stimulus situation at the time the subject responds is always the same, yet the subject correctly makes the two different responses required by the problem. French arrived at his conclusion after attempting to analyze alternation problems according to the S–R interpretation of animal memory. However, there simply are not any differential stimuli to elicit the two different responses. At one time (Osgood, 1953) it was thought that motor cues were differential stimuli, in the sense that turning in one direction on trial n is the signal to turn in the opposite direction on trial $n + 1$. However, as French points out, this explanation is totally inadequate because rats can carry out delayed spatial alternation even though they have been anesthetized during the intertrial interval (Ladieu, 1944; Loucks, 1931) and rhesus monkeys are capable of nonspatial delayed alternation in which any orienting response would be useless (Behar, 1961; Pribram & Mishkin, 1956).

An alternative way to explain successful alternation performance in animals is to assume that they hold a representation of the information supplied by their response on trial n in memory and then use that information as a basis for their response on trial $n + 1$. This explanation does not indicate the nature of the information held in memory but it begins to direct attention toward asking this question.

Delayed Matching to Sample

In the usual delayed matching to sample (DMTS) paradigm a sample stimulus is presented, a delay follows in which no stimulus is present, and then two stimuli are presented. One of these two stimuli is identical to the sample stimulus and the other is different. The subject's task is to respond to the stimulus that is identical with the sample stimulus. Performance in the DMTS task may be explained in terms of S–R memory, because there is a stimulus present (the sample) to elicit the correct response. However, D'Amato and Worsham (1974) have developed a variation of this task, referred to as "conditional matching" (as opposed to the traditional DMTS task, which involves identity matching), which seems to involve representational memory. The task requires the subject to "match" on the basis of an arbitrary association instead of identity. For example, a sample, such as a red disk, was presented and then, following a delay, the subject was presented a triangle and a circle and required to pick the triangle. If the sample was a vertical line the subject was required to pick the circle. Thus, at the time of response the stimulus situation was always the same and the subject was required to make different responses (picking the triangle or the circle), based on information provided by the sample that was not present during the delay or at the time of the response. In order to respond correctly in this situation the subject would have to recall the information provided by the sample and then use that information to determine his response. The capuchin (*Cebus apella*) monkeys performed well above chance in this situation over delays of up to 100 sec. Their performance cannot be explained in terms of S–R memory.

Konorski (1959) has proposed a test of recent memory, derived from the DMTS task, that is well suited for testing representational memory in animals. In his method, two stimuli are presented successively. Depending on whether the two stimuli are alike or different from one another, dissimilar responses are required. Stepien and Cordeau (1960) first applied Konorski's paradigm to nonhuman primates. The stimuli they used were 2-sec series of evenly spaced clicks (either 5, 10, or 20 per sec). A series of training steps were used but by the end of the study all possible two-stimulus combinations were presented in random order with a 5-sec delay between them. The monkeys were required to open the door to a food box if the two stimuli were the same and to not open the box if they were different. Their performance was well above chance. In a final phase of the experiment, flashing lights were substituted for the clicks and again the monkeys performed well above chance.

Recently D'Amato and Worsham (1974) modified Konorski's method along the lines suggested by Stepien and Cordeau in their paper. Specifically, D'Amato and Worsham required different responses to indicate that the second stimulus was the same (key press) or different (lever press) from the first stimulus. This eliminated the difficulty encountered by Stepien and Cordeau in training animals to withhold responses in their go, no-go procedure. Instead of using temporally patterned stimuli they used a red disk, a vertical line, a square, and a triangle. They reported that their capuchin monkeys performed very well on the task at a level that was comparable to their delayed matching to sample performance for delays of up to 16 sec between the first and second stimulus. At longer delays (32–128 sec), performance dropped some but was still well above chance and still comparable to their performance in the delayed matching to sample paradigm.

These studies provide an excellent demonstration of the ability of animals to perform well in a task that requires representational memory. The subjects had to make different responses when faced with a stimulus situation that could in no way indicate in and of itself which response was correct. Instead, they had to base their response on a stimulus that was not present at the time of the response. Clearly, this requires that some kind of representation for the information provided by the first stimulus be held in memory for delays of up to 128 sec. Another interesting feature of Konorski's paradigm is that it not only requires the subject to respond on the basis of a stimulus that is not present, but it also requires him to use that information in conjunction with information presented later in order to respond correctly. This characteristic of Konorski's paradigm would seem to make it ideally suited to studying the relative rates of decay or distortion of various stimulus dimensions or attributes. To our knowledge, this has not been done.

The Nature of Representational Memory

Until now we have simply gathered together what seems to us to be evidence that animals do have some kind of representational memory, but we have not asked what this representational memory is like. If Underwood were to write the animal analog of his "Attributes of Memory" paper (Underwood, 1969), he would indeed have little to say. As a start the studies mentioned so far indicate that animals are capable of holding a representation for form and color information (D'Amato & Worsham, 1974); the nature of rewards (Tinklepaugh, 1928; Medin's study cited in Medin & Davis, 1974); spatial position of a stimulus (Ruggiero's MPR studies); alternation responses (French, 1965); and temporally patterned stimuli (Stepien & Çordeau, 1960). Despite these results, little is known concerning the exact nature of these representations, for instance the manner of their storage and subsequent retrieval.

The study by D'Amato and Worsham (1974) illustrates this problem. They reported that their monkeys performed equally well in delayed matching to sample and delayed conditional matching tasks. In the first case many obvious retrieval cues are present because the sample itself is present at the time of the response. However, in the conditional matching task many retrieval cues are eliminated because the sample stimulus is not present at the time of the response. The fact that the monkeys perform equally well in both tasks conflicts with the evidence presented in the human memory literature, where recognition tasks almost always result in better performance than recall tasks. More research on retrieval of representational memories in animals is needed to clarify this picture.

Transformation of memory trace in monkeys. One way to discover what a subject's memory for an item is like is to determine what the subject can do with his memory for that item. Piaget's work (Piaget & Inhelder, 1973) with children is a good example of this approach. He used the ability to perform certain operations as the key to understanding various cognitive processes. This strategy was used in the next two experiments to investigate further monkey memory employing the refinements of the MPR paradigm developed in the view, no-view study. We asked whether animals could use information held in representational memory, in combination with new information, to determine a totally new response not indicated by either set of information alone. This was accomplished by requiring the monkeys to perform a specified operation on information held in memory. That is, the subject not only had to remember which cell of the matrix had been illuminated (stimulus cell) but also had to perform an operation on this information in order to arrive at a correct response to a new cell (response cell).

In the first of this series of studies (Flagg, 1975), the stimulus cell was embedded in a colored row. All of the cells in the row containing the stimulus cell were illuminated one color (original background); then one cell (stimulus) in the colored row was illuminated with white light for a brief time. Finally, the stimulus light and colored background row were extinguished and a different row was illuminated (new background). The correct response was to the same left–right position in the new background row that the stimulus had occupied in the original background row. The subject was therefore required to remember the stimulus information and to combine it with the new background information according to a specific operation in order to arrive at the correct response.

On a typical trial the following sequence of events took place (see Fig. 1.3). The experimenter raised the opaque door of the WGTA and one row of the matrix was immediately illuminated with one of the four possible colors of light (original background). One second later, one of the cells in the colored row was illuminated with white light for 0.5 sec and then returned to the color of the rest of the cells in the row. The row of colored lights remained on for 1.0 sec after the white stimulus light terminated; then the background was also extinguished. After a 0.5-sec delay to prevent masking effects (Harris, 1970), either the same row or one of the other three rows was illuminated the same color as the original background. Immediately following the appearance of the new background, the experimenter raised the viewing screen, allowing the animal to respond once. In order to facilitate learning, each problem was presented four times before the next problem was begun.

The relationship between the original background row (the row that contained the stimulus) and the new background row was that they occupied either the same position, a position one row apart, a position two rows apart, or a position three rows apart. The percentage of correct responses on all four trials within a problem is shown, as a function of 3-day blocks of practice, in Fig. 1.4 for each of these possibilities. The condition in which the two backgrounds occupied the same position served as a control condition because the stimulus and response cells were identical and no transformation was required. As can be seen from Fig. 1.4, the performance of the animals was at a high level on this condition throughout the experiment. The other three conditions required the monkeys to

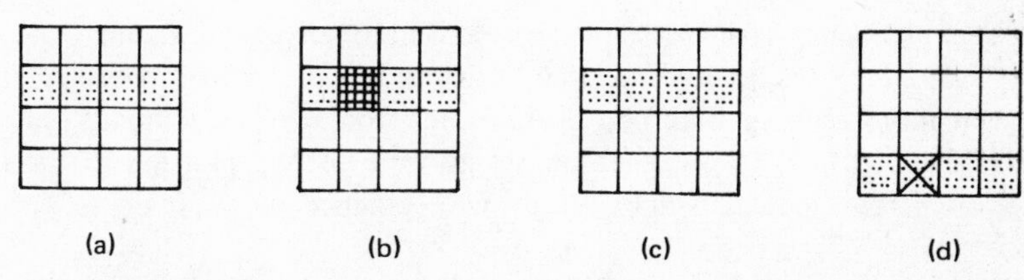

FIG. 1.3 Stimulus display conditions for the basic transformation of memory task. One of the rows is illuminated a color (a: original background), and then the stimulus cell (crosshatched cell in b) is illuminated with white light. The stimulus light extinguishes and the original background remains lit (c). Finally another row is illuminated the same color as the original background (d: new background), and the animal is allowed to respond to one of the cells. The correct response is signified by the X, but this cell is not illuminated at the time the animal makes its response.

perform an operation on the stimulus information in order to arrive at a correct response. The practice effect in these three conditions indicated that the monkeys did learn the required operation and were able to apply it, although not with equal ease for all conditions.

The types of errors were examined in order to determine how the monkeys might have learned the operation. The majority of the errors can be grouped into responses to the original position of the white stimulus light and responses to an incorrect position in the new background row. The first category of errors accounted for 25% of all errors early in training and 5% of all errors by the end

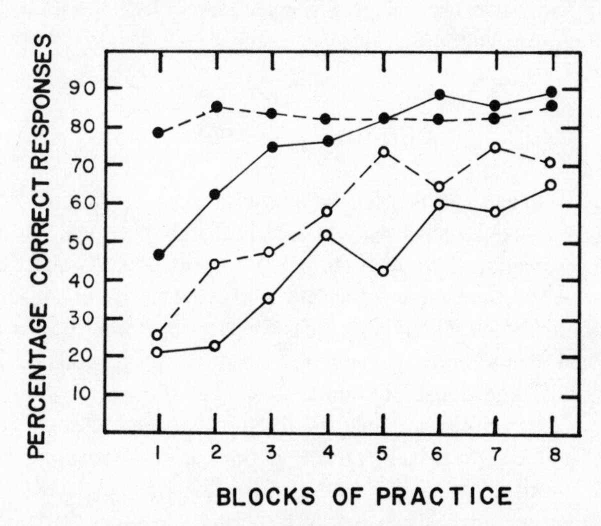

FIG. 1.4 The percentage of correct responses made in the basic transformation of memory task, as a function of 3-day blocks of practise and the amount of displacement for the new background row from the original background row. ●– – –●: Same background row; ●——●: one row away; ○– – –○: two rows away; ○——○: three rows away.

of the experiment. The relatively large percentage of errors in this category, early in training, was no doubt related to the prior experience of the subjects on problems in which they were reinforced for responding to the stimulus cell. The other type of error, involving responses to an incorrect position in the new background row, increased dramatically with practice from 36 to 83% of the total errors.

Performance on trials where the new background was displaced by one row from the original background indicated that the animals learned the required operation very well by the end of the experiment. If they learned the rule, however, why was performance somewhat poorer when the new background row was displaced two or three rows from the original background? One possibility is that the appearance of the new background interfered with their memory for the stimulus position. However, this is unlikely because when the new background was displaced from the original background by one row, performance was as high as in the control condition. A more likely alternative was that, although they had learned the operation by the end of the experiment, they had trouble applying it. That is, when the new background was displaced from the original background by two or three rows it was more difficult to judge which cell in the new background row occupied the same position as the stimulus had in the old background row. It should be noted that a second experiment demonstrated that monkeys can perform well in this task when a delay of up to 12 sec is imposed between the offset of the original background and the onset of the new background. In addition, eliminating the animal's view of the matrix during the delay period had no effect on performance. Overall, these studies indicate a remarkable flexibility in monkeys.

ORGANIZED MEMORY

The fact that animals seem to hold a representation of stimulus information in memory, one that can be retrieved and used in the absence of the stimulus itself, leads to two interesting questions. Perhaps the most obvious question concerns whether representational memory shows any evidence of organizational processes. A second but closely related question concerns whether or not several representational memories can be combined to form general information, or what we may call "knowledge" in humans.

Tulving (1972) proposed a similar distinction between episodic memory and semantic memory in human subjects. He defines *episodic memory* as a memory for a specific event that has occurred at a specific time. What we have labeled "representational memory" is analogous to episodic memory. Tulving goes on to define *semantic memory* as general knowledge derived from more than one episodic memory. For an example, a person may have several memories stored in episodic memory that consist of seeing birds fly on several occasions. From these separate memories it is possible to arrive at a piece of general knowledge consisting of the fact "birds are capable of flying." This general knowledge is stored in semantic memory.

If animals are capable of storing information in representational memory that can be retrieved and used in the absence of that stimulus, can they use this information to generate a new piece of information that has not been contained in any one of the separate representational memories? In other words, are they capable of acquiring what may be called "general knowledge" or "semantic memory" in humans? Because the term "semantic memory" is closely tied to language in humans, we shall use the term "organized memory" instead when referring to this kind of memory in animals.

MPR Studies

A number of studies have used the basic MPR paradigm and found evidence for organized memory in monkeys (see Medin & Davis, 1974, and Davis, 1974, for a review). In one such study, reported by Medin (1969), rhesus monkeys were tested on their ability to remember four-light patterns in the MPR task. Increasing the complexity (number of sides) of a four-light pattern resulted in a decrement in performance, and this effect is not an artifact of response biases nor is it related to processing and responding time (Davis & Ruggiero, 1973).

The probability of remembering a four-light pattern cannot be accurately predicted on the basis of performance of the constituent four cells when they are individually presented as one-light patterns. Instead, the memory for the four-light pattern is organized and more complex patterns are more difficult to organize and remember.

A related study by Motiff (1969) showed that horizontal patterns composed of the four cells in a given row are easier for monkeys to learn and then relearn, after 24 hr, than either vertical (columns) or diagonal four-light patterns. In both the two- and the four-light cases, diagonal separations and diagonal patterns were harder to reproduce than their horizontal or vertical counterparts. These results suggest that monkeys do not encode both horizontal and vertical patterns as straight line patterns and indicates that monkeys have trouble with diagonal patterns, as do children (Rudel & Teuber, 1963), chimpanzees (LeVere, 1966), rats (Lashley, 1938), and octopus (Sutherland, 1957).

Finally, symmetrical four-light patterns (cells that are corresponding distances apart from the central axis) are no more easily reproduced than asymmetrical patterns (Medin, 1969, Experiment 4), a finding contrary to that in studies with human subjects (Attneave, 1955).

Organized memory is not necessarily restricted to primates. Hertz (cited in Ellis, 1955) tested the organized memory of jay birds (*Garrulus glandarius*) over a series of complex, direct delayed-response tasks and concluded that the organization of their perceptual memory "... is essentially the same as our own" (p. 252).

Learning Set

Organized memory can also be obtained from an examination of the learning-set situation (Harlow, 1949). A typical object-quality, learning-set procedure consists of presenting animals with a large number (several hundred) of discrimination problems. Each problem consists of a limited number of trials (two to six, usually) with two distinct stimuli that are often common dime store objects (common use and manufactured objects or "junk" objects) which vary on a number of dimensions. The two objects used for each problem are unique and the subject is required to respond to an arbitrarily selected member of that pair in order to receive a reward. Because the left–right position of the correct object is randomized and other differential cues are eliminated, the subject must learn to consistently choose the correct object in order to consistently receive reinforcement. Animals typically do relatively poorly on the first problems. However, their performance gradually improves over problems until they eventually reach a level where they are almost always correct on every trial of a given problem after Trial 1. (Performance on Trial 1 must remain at the chance level because the subject has no way of knowing which stimulus is correct until he has responded once.) When animals reach this stage they are termed "learning-set sophisticated."

This is a nice descriptive term, but what is involved in becoming learning-set sophisticated? The answer depends in part on one's theory of learning-set formation, but even on the empirical level something more than episodic memory must be involved. For example, the animal must learn that a Trial 1 reward is to be associated with the object rather than the position. Riopelle and Chinn (1961) trained monkeys on a procedure in which the reward on Trial 1 of new problems was always in one position (say left) but thereafter reward was associated with the correct object regardless of its position. In other words, Trial 1 performance must be controlled by positional responding, yet the monkey must encode the outcome (reward or nonreward) in relation to the objects, not the positions. Monkeys were able to perform above 80% correct on Trial 1 of new problems and were above 90% correct on the second trials, even though the correct object was in a different position on half of such trials. Clearly, animals can store information gathered over a number of trials and problems and then combine it to arrive at an appropriate strategy. The strategy cannot be derived from the information obtained on any one trial. A large number of particular experiences must be integrated to evolve strategies and direct behavior. We suggest that the acquisition of these strategies can most parsimoniously be thought of as involving organized memory.

CONCLUSION

We have gone to considerable effort to distinguish S–R, representational, and organized memory. Our aim is to provide a means of organizing animal memory research that for us at least, can generate a large number of questions that may not otherwise have been considered. Concern with what we have called S–R memory focuses questions on the relationship between learning, interference, and forgetting; concern with representational memory immediately raises questions about the attributes of memory and their relation to the retrieval of information; finally, concern with organized memory gives rise to questions "mostly unasked and unanswered" related to the integration of a set of experiences. We need answers to many of these questions in order to arrive at a better understanding of animal memory.

Finally, we can return to the philosopher in our anecdote and begin to answer the question, do animals have memory? For monkeys at least (one can speculate about simpler organisms, such as planaria) the answer is yes. Yes, in all three aspects we have outlined. The significance of this answer is that we have no basis for assuming major discontinuities between animal and human memory. This conclusion supports our belief that a more complete understanding of human memory processes can effectively derive from an integration of human and animal memory research.

ACKNOWLEDGMENTS

The original research reported in this chapter was based on dissertations submitted by both authors in partial fulfillment of the requirements for the Doctor of Philosophy Degree at Washington State University, and was supported by United States Public Health Service Grants MH19321-03 and HD05902-03 to Roger T. Davis, Comparative Behavior Lab, Washington State University.

2

Some Relationships between Learning and Memory

Robert C. Bolles

University of Washington

INTRODUCTION

If an experimental situation provides an ordering of events such that one event follows another in a lawful manner, and if an animal's behavior changes as a result of this arbitrary ordering, then we may say learning has occurred. The learning itself is difficult to identify because it is not an observable thing or relationship but is a hypothetical process, which has been variously conceived as forming a neural connection, building a habit, establishing an association, or acquiring an expectancy. But however the process is conceived, the change in behavior merely constitutes evidence of learning. Nonetheless, when an animal's behavior changes so as to reflect the ordering of events, this evidence may be interpreted as indicating that the animal has learned the order. In general, then, learning may be defined as the process of associating ordered events. Note that there is no evidence of learning at the time learning occurs; evidence of learning consists of a subsequent change in behavior. Therefore any test for learning must necessarily also be a memory test.

There are other logical relationships between learning and memory. One obvious relationship is that there must be memory if there is to be any learning. If an animal is to learn to associate two events, then the first event must be remembered in some way and in some form until the second event occurs. If the two events are close enough together in time, then some sensory storage mechanism or other short-term memory device may suffice. However, there is now a good deal of evidence to show that animals can remember prior events over substantial periods of time and associate them with subsequent events, and these cases require us to think of a relatively long-term memory mechanism. Let us adopt Revusky's (1971) label "associative memory" to refer to this necessary memory function, that is, the necessity to remember a prior event if it is to be associated with a second event occurring some time later.

If an animal is to demonstrate the cumulative effect of repeated training trials, or of repeated training sessions, it must be able to remember what is learned from one trial to the next, or from one session to the next. Without some long-term memory of what is learned, which Revusky (1971) calls "retentive memory," there would be little learning evident in the animal kingdom and there would be little reason for animals being able to learn. Although there are numerous examples of one-trial learning in the animal literature, there seem to be no examples of learning reaching an asymptote in one trial. Learning as it is commonly seen in animals therefore depends on animals' having both associative memory and retentive memory capacity.

It should also be noted that were it not for learning, that is, for acquired associations between events, animals would have nothing to remember. It is hard to see what part memory might play in fixed behaviors of the instinctive or reflexive variety, or even how memory might accompany such behaviors. Perhaps it was this natural dissociation of fixed and automatic behavior, on the one hand, and memory, with its cognitive implications, on the other hand, that led the early learning theorists to put learning firmly on the side of automatic behavior (i.e., the learned S–R connection was said to be like a fixed S–R connection) and then pay so little attention to memory and related phenomena. Whatever the historical reasons for it may have been, one of the most interesting relationships between learning and memory is that all of the classical animal learning theorists have had little or nothing to say about memory. Some (e.g., Guthrie, 1935) denied that memory was a viable explanatory principle; Guthrie accepted no associative memory mechanisms. For him, learning depended on the strict contiguity of stimulus and response. Hull (1943) had a short-term memorylike factor in the "stimulus trace" but, again, learning involved attaching a response to the fading trace rather than to any memorial representation of the original stimulus. For Hull, therefore, there was no real associative memory mechanism per se; learning depended on the simultaneous occurrence of the response and whatever was left of the stimulus. Spence (1947) not only endorsed the principle of contiguity but added to it by proposing that reinforcement also had to occur almost immediately after the stimulus and the response if learning was to occur. I will have occasion to discuss this proposal in more detail later.

The early S–R learning theorists also either discounted or denied retentive memory phenomena. Pavlov, Guthrie, Hull–the whole tradition of S–R theorists–had no place for common forgetting. Forgetting was inevitable, like death and taxes, and the learning theorist could not deny its existence. However, they could and did deny it systematic treatment. The old studies showing that less forgetting occurred during sleep than during an equal period of wakefulness were eagerly seized on as demonstrating that there really was no such thing as forgetting, in the sense of loss of storage. There was only interference produced by new learning. Memory and forgetting were not basic phenomena, therefore; they were only secondary effects to be derived from the principles of learning. Learning was the fundamental domain and perfect retention was to be expected except in cases where counterconditioning or some other type of competing learning had occurred. Gleitman (1971) cites a number of early experiments showing retention over very long intervals. Evidently what happened in these cases was that the experimenter discovered an old, misplaced animal that had been a subject in a discrimination study the year before. When the animal was put back in the old apparatus it still responded correctly. Gleitman observes that when the learning theorist describes these results, he characteristically emphasizes that there was some retention over the long interval and rarely ever notices the forgetting. Actually, this conceptual bias was only part of the story. Another important part of the picture was that the psychologists who were primarily interested in learning phenomena rarely carried out the experiments that would have displayed memory effects. As long as interstimulus intervals and intersessions intervals were not too long, there would be some retention, and the psychologist could proceed with his study of learning. Of course, moreover, the long history of emphasis on learning phenomena produced an evolution of experimental procedures that made it increasingly easy to demonstrate learning phenomena and increasingly difficult to show memory and forgetting effects. Indeed, the burden has been on the psychologist interested in memory to demonstrate that his behavioral effects are not caused just by familiar learning processes, and he is obliged to invent entirely new procedures to convince us that he really is assessing memory. Gleitman's (1971) report of his own ingenious research designs illustrates the point.

There is another perspective from which to view the relationship between learning research and memory research. There are several behavioral effects that are well known to the community of learning psychologists, and although these may be viewed as memory phenomena, they are most often regarded as special effects lying entirely within the province of the learning theorist. One example is the so-called "warmup" effect. The origins of this term seem to be lost in the obscurity of the general culture, but it suggests that when an organism returns to a familiar task, it may be expected to perform poorly until it gets back "into" the task again. The decrement in performance is variously attributed to a temporary loss of motivation (for example, there is a common assumption that warmup effects in avoidance behavior are caused by the dissipation of fear during the intersession interval), or to the necessity of recapturing the precise stimulus conditions under which the behavior was previously established (Spear, 1973). Warmup is seen as a temporary loss of stimulus or motivational control. The implicit idea is that the animal has to get back the postural stance, the orienting behavior, etc., which it had at the end of the previous session. Ideally, if the entire situation is the same as before, then the behavior should be the same as before. This assumption is hardly subject to empirical confirmation because of the difficulty of objectifying proprioceptive feedback stimuli, orienting behaviors, and learned sources of motivation. Still, this is the customary interpretation of the warmup effect among learning theorists. Although there can hardly be any clearer example of forgetting over the intersession interval and relearning during a session, it is only recently that theorists, such as Spear (1973), have begun to view the warmup phenomenon as a memory effect.

Another example of a learning interpretation of a memory phenomenon has already been noted; namely, the interference view of the behavior decrements that occur over time. Loss of behavior over time has generally been viewed by learning psychologists as a result of new learning that displaces the old learning, rather than as forgetting.[1] A third example is the importance attached in recent years to attentional processes in animal behavior. Sutherland and Mackintosh (1971) have described a great variety of phenomena that now tend to be interpreted in terms of selective attention. If an animal performs poorly on some task, the reason is said to be its failure to attend to the appropriate dimension of the stimulus. Again, this loss is not attributed to forgetting but to the failure of a new variety of learning, attentional learning.

One of the most curious relationships between learning and memory is that the trade of ideas and concepts between these two areas has occurred almost entirely in one direction. The students of learning have, so far, gained very little from the work of their colleagues studying memory. By contrast, much of the conceptual machinery that has been used to explain memory phenomena has been adopted from concepts originally developed in the animal learning area. In fact, some of the behavioral concepts developed by learning theorists in the attempt to discount memory as a legitimate area of investigation have been taken over to provide systematic accounts of memory phenomena. Each of the learning phenomena cited in the previous paragraph has been incorporated in one way or another into the systematic analysis of memory and forgetting. One striking instance was McGeoch's (1932) hypothesis that forgetting was caused by the learning of competing response tendencies. The original S–R association was assumed to be intact, but the S no longer evoked the R because a new, competing R had become associated with it. Forgetting was therefore said to be caused, not by a loss from storage, but by a disruption of what was stored. This

[1] There are two possible sources of confusion in this statement. One is that "forgetting" is often defined operationally as any loss of associative strength with time, including the losses produced by new learning. Interference therefore becomes a source of forgetting because it produces response decrements. My statements reflect my own view that "forgetting" should mean not just a loss that occurs over time, but a loss from storage that occurs because of the lapse of time. I see the term as the name for a hypothetical process, such as the decay theory proposed. Interference therefore becomes an alternative to forgetting as an explanatory mechanism to account for certain kinds of behavior decrements. The second source of confusion emerges in the next paragraph; it is that memory theorists have tended to use the very mechanism by which learning theorists bypassed forgetting to explain forgetting.

general approach has been advocated by Underwood (1957) and Underwood and Postman (1960) as a scheme to explain a great array of memory phenomena. A second example of this one-way trade is the work of Tulving (1972), showing the importance of "retrieval cues" in alleviating memory loss. It is not that material is lost from storage so much as that it cannot be gotten out. Retrieval cues effect retrieval by reinstating the conditions under which the material has originally been put into storage. This general argument is quite familiar; we have already encountered it in connection with the warmup effect. A third example is the notion, well documented by Melton and Martin (1972), that failures to remember can often be attributed to inadequate input to storage at the time of original learning. A variety of recent human learning studies pointing to the importance of rehearsal help substantiate the idea that later performance depends on appropriate initial attention to the material to be learned. This idea is remarkably similar to Sutherland and Mackintosh's (1971) analysis of attention in animals. If an animal fails to attend to the appropriate aspect of the situation, it will evidence little learning. Sutherland and Mackintosh's analysis of attention in animals provides a new advance in our thinking about such matters, but the idea that orienting responses make important contributions to behavior is an old tradition in learning theory. The idea, expressed in S–R terms, dates back to Spence and his students (e.g., Ehrenfreund, 1954) and before that to Guthrie and Pavlov. Oversimplifying a bit to summarize, we can say that input functions in memory derive historically from the concept of orienting behaviors in animals, storage functions derive from the concept of counterconditioning, and retrieval functions derive from the explanation of the warmup effect.

Perhaps more than anything else, students of memory took over from the learning theorists the whole S–R approach to behavior analysis. This heritage, fortunately, has been largely overcome in recent years. It is clear, however, that although memory psychologists have made considerable use of concepts and principles from learning theory, they have also had to carve out their own territory and then to defend it against the learning theorist. My own view of the matter is that the time is past due when the learning psychologist should start paying attention to what his colleagues in the memory area are doing. The investigation of memory has produced many conceptual tools that could be of great value in working out a better understanding of learning in particular and behavior in general.

Two kinds of recent events indicate the time has come for the establishment of a new, more productive relationship between learning and memory. One development is that it is no longer possible for the learning theorist to ignore memory phenomena or to deny their importance for his own discipline. There is also an increasing number of curious phenomena (curious from the perspective of conventional learning theory) that have been discovered in recent years that only seem to make sense when interpreted as memory effects. Some of these phenomena are discussed in the next section. The second development is that the traditional S–R approach to behavior, which originally gave rise to much of what now constitutes the areas of learning and memory, no longer seems very fruitful. Learning theory is currently undergoing a revolutionary change (my view of which is described elsewhere; Bolles, 1975a), and it is therefore particularly receptive to new concepts and ideas.

One additional introductory note: Learning is characterized here as an acquired association between events. Let me be more specific, and more personal. I have tried to develop the idea (Bolles, 1975a, b) that learning occurs in two forms. Neither form is S–R. One is S–S*, where S stands for a predictive cue, that is, a cue that is correlated with the occurrence of some biologically important stimulus event, S*. This kind of learning is most easily demonstrated in a Pavlovian situation where the correlation between S and S* is strictly controlled by the experimenter. Evidence for S–S* learning is found when the animal is presented with S alone and behaves as though it expects S*. The second form of learning, I propose, is found in an operant situation where there is a contingency between some specified response, R, and S*. Then R–S* learning may or may not occur; its existence is often difficult to be sure of because the animal's behavior may be controlled by S–S* contingencies embedded in the experimental situation. What the animal must remember, then, if it is to show learning, is the relationship between S and S* and/or the relationship between R and S*. An examination of some memory phenomena should provide evidence for or against this view of learning.

MEMORY AND COGNITION

In the great majority of learning situations there is no immediate way to know whether observed changes in behavior are more properly attributed to memory mechanisms or to learning mechanisms. For example, suppose a rat is repeatedly reinforced by food for running in an alley. The rat's progressive increase in running speed could conceivably be attributed to its gradually improved memory of food at the end of the alley, or to an improved memory of the spatial layout of the apparatus–where food is located–or to an improved memory of what response gets it to the food, or to some combination of such memory factors. In short, faster running could be attributed to a combination of better memories or more accurate memories of the S–S* and R–S* relationships in the situation. However, such an interpretation seems strangely alien and unnecessarily complicated to the animal psychologist. The trouble is that we have become so accustomed to the conventional learning theory explanation of such effects: the food reinforcement produces a gradual strengthening of the "connection" between the running response and environmental stimuli. It should not be surprising that the learning interpretation fits this kind of situation so comfortably; it was originally derived from just such observations.

However, there has always been a small fringe of studies to which the learning interpretation cannot be applied so comfortably. These are mainly experimental situations in which there is no opportunity for the appropriate response to become connected with the right stimulus pattern so the animal is obliged to behave "constructively" on the basis of its prior experience. The latent learning experiments illustrate the point. The rat is permitted to explore the apparatus, but no food is present. Then immediately after food is introduced the rat executes its passage through the maze (Tolman & Honzik, 1930a) or through the T maze (Seward, 1949) nearly as well as control animals for which the appropriate response has been previously reinforced. The most parsimonious interpretation of these results would seem to be that the rat remembers where the food is, remembers how to get there, and, putting these pieces of information together, goes to the food place. The S–R theorist can explain the results too (e.g., Hull, 1952), but to do so he must hypothesize a special class of unobserved responses (r_G), give them special unobservable properties (feedback, s_G), and make further special assumptions about sources of reinforcement in the situation other than food.

The "insight" experiments constitute another problem for the S–R learning position. Here, the rat has learned to run for food, but one day the path is blocked and the rat must find a detour (Maier, 1931; Tolman & Honzik, 1930b). Again the question is whether the rat will behave "constructively" by using information about the spatial arrangement of the apparatus (i.e., its "cognitive map"). If it does, we may say that it remembers different pathways, remembers where food is, and that it integrates these pieces of information about the environment to solve the problem. While the S–R learning theorist has some purchase on the latent learning phenomenon (because he can emphasize that the correct response has occurred before and only needs to be properly motivated to recur), he cannot deal effectively with the insight studies because of the constructive nature of the new response. A number of further experiments carried out by Tolman and his students could be cited, which would provide more evidence against the conventional S–R interpretation. Of course, Tolman's work was largely dedicated to refuting the conventional position, so this conclusion is not surprising. What is surprising, perhaps, is how readily much of the research from Tolman's laboratory lends itself to an analysis in terms of memory. There seems to be a natural affinity between the cognitive approach to behavior and the study of memory, perhaps only because they have a common opponent in the traditional S–R psychology. There must be more to it, however. If the rat is to behave cognitively, as Tolman supposed, then it must make use of stored information, and information about environmental relationships must be remembered before it can be expressed in behavior. Memory is a necessary part of cognitive theory. It is probably not just a coincidence that Gleitman, who has played such an important part in animal memory research, was one of Tolman's last students.

One difficulty with Tolman's cognitive view of behavior was that he tended to see no limit to the amount and complexity of information the rat could handle, no limit to what it could remember, and no limit to the ingenuity with which it could construct new behaviors. He had great faith in the rat's cognitive ability, and I suspect that he felt a kind of personal loss whenever the literature reported a new case of rats not showing latent learning. However, it is now generally recognized that where there is cognition there must also be limitations on cognition. Where there is memory there must also be forgetting. It would seem foolhardy to maintain today that the rat, or any other animal, could store whatever information the environment provided. Memory is selective. Even humans, who can process all sorts in imput, are selective in what they store. Humans show enormous individual differences as well as great species specificity in the kinds of material they remember. Our task then, as I see it, is to discover in what ways the memory of a particular kind of animal is selective. What kinds of information does it retain, and what kinds of information are either never processed or soon lost? I will keep returning to this question in the following sections.

There is a second, related question that should be kept in mind as we proceed: In what form does an animal store information? To illustrate the question, consider a rat in a Pavlovian fear-conditioning situation. After a number of pairings of a tone with shock, the presentation of tone alone produces a change in heart rate. What specifically has the animal remembered? There are several options. One possible answer is that nothing has been or need be remembered; a direct connection has been formed between tone and heart rate and no further psychological devices, such as memory, are required to explain the new behavior. It is clear that Pavlov (1927) deliberately coined the term "conditioned reflex" to emphasize the automatic, physiological nature of such new behaviors and to obviate any psychological interpretation of them. This was a provocative and productive view of the matter, but recent reviews of conditioning (e.g., Bolles, 1975b; Kamin, 1969; Rescorla, 1972) have revealed that there is now evidence for a variety of psychological effects that occur in fear conditioning. We may expect that memory is one of them.

A second, conventional approach is to suppose that conditioning consists of forming a new S–R connection, and that this connection is subject to forgetting. It is the S–R connection itself, therefore, that is remembered or forgotten. There is abundant evidence from studies of retentive memory in animals to support this position; behavior usually does weaken over a retention interval. And, of course, this kind of interpretation has enjoyed a long and distinguished tradition in human learning; it began with Thorndike and has been brought up to the present day by a number of theorists, including Postman and Underwood. The S–R analysis has always seemed somewhat artificial when applied to human learning, however. One word or nonsense syllable was designated by the experimenter to be the stimulus and another was designated to be the response. A mass of recent research on human memory has shown that this kind of paradigm just scratches the surface of human information processing. Words are basically words, it seems, and human subjects encode, retain, and retrieve them by using all the meanings, groupings, and other devices the situation permits. Similarly, in the world of animal learning, there are now many reasons for doubting that learning must necessarily consist of S–R connections. There is therefore little reason to think that the contents of memory must necessarily have the S–R form.

To return to the frightened rat: perhaps it remembers the S–S* relationship. That is, perhaps it remembers the specific correlation between tone and shock that the experimenter has imposed on it. Some evidence for this view is described shortly. A different possibility is that the contents of memory are discrete events, rather than relationships between events. Perhaps the rat merely remembers S*, and S serves primarily as a retrieval cue to get the memory of shock out of storage. Such a mechanism would account for why the cue could produce such a well-defined and temporally located expectancy of shock.

In the operant situation there are both S–S* and R–S* contingencies, and there is a corresponding increase in the number of logically possible contents of memory. This analysis does not need to be carried out here, but it should be borne in mind that there are a number of possibilities. It is to be hoped that, as we discover more about animal memory, it will be possible to select meaningfully from among these possibilities and obtain a better understanding of the structure of animal memory.

SOME MEMORY PHENOMENA

One of the oldest series of studies addressed specifically to the question of animal memory is that dealing with alternation behavior (Munn, 1950, gives a good account of the original studies). The early experimenters, who were not as committed as later generations to the S–R position, were concerned about memory. Can the rat remember where it has gone on the previous trial and use that memory to consistently make a different response? Hunter (1929) convinced himself that the simple alternation procedure was methodologically inadequate, and he devised a double-alternation procedure in which the rat had to turn LLRRLL, etc. To solve this the rat must remember the last response and the response before that, and must then solve all the subproblems, that is, having gone LL is a cue to turn R, having gone LR is also a cue to turn R, but RR is a cue to turn L, etc. A logically simpler but psychologically more taxing strategy for the rat would be to ignore the memory of the last response and use only the memory of the response before it to make the opposite turn (talk about retroactive interference!). Hunter's stringent memory test proved to be too much for the rat. Nor have subsequent investigators obtained double alternation in animals except in those situations where each choice point is distinctive or where the intertrial interval is very short so that there can be "chunking" of responses. There evidently are limits to the amount of information the rat can handle in this type of situation.[2]

In spite of this unpromising beginning, the alternation paradigm has provided substantial evidence of animal memory. Petrinovich and Bolles (1957) had found considerable spontaneous alternation in rats trained to go to one side of a T maze. This behavior was a nuisance, but we capitalized on it by reinforcing other animals for alternating (Hunter, too, had no trouble training simple alternation). We found that alternation was learnable, and that when the rat had learned to alternate, it could do so when the intertrial interval was increased to 2 or 3 hr. The best indication that alternation was dependent on memory was that it ultimately broke down in all rats when the intertrial interval was lengthened to 6 or 8 hr. The existence of alternation with well-distributed trials indicated that it was not caused by any inhibition mechanism, and other aspects of the data made an S–R interpretation of alternation untenable. Petrinovich and I therefore drew the only conclusion we could think of: The rats remembered where they had gone on the previous trial and went to the opposite side because they were rewarded for doing so.

John Capaldi is responsible for much of the subsequent development of this theme. Cogan and Capaldi (1961) found that rats rewarded and not rewarded on alternate trials soon showed anticipatory alternation of running speed–fast on rewarded trials and slow on nonrewarded trials (Crum, Brown, & Bitterman, 1951, had previously reported this effect). At the same time Cogan and Capaldi found no such anticipation when they alternated immediate and 20-sec delayed reward. The two sets of conditions provide some control for the possibility that the effects result from frustration or other conventional S–R mechanisms. We are left with the conclusion that rats can remember whether they have been rewarded on the last trial, and that this memory can control the running response on the next trial. An interesting feature of Capaldi's results is that the memory of the previous trial evidently serves a purely informational function; or if there is an incentive motivation effect here, it is masked by the information effect. Thus, it might be supposed that when the rat remembers being fed on the last trial, this memory would motivate faster running on the next (nonrewarded) trial, but the alternating pattern of running speeds does not follow this pattern.

Capaldi has continued to investigate this phenomenon and has applied it to the explanation of several vexing animal learning problems, the most notable of which is the partial reinforcement extinction effect (Capaldi, 1967). The gist of his argument is that with partial reinforcement there are occasions on which the

[2] There were other early experimenters who were interested in memory. For example, Warner (1932) varied the CS–US interval across groups in one of the very first avoidance learning experiments, to test associative memory in rats. However, all the animals performed so poorly that nothing could be concluded about associative memory.

animal runs while remembering nonreward on the previous trial (N) and is then rewarded (R). Then, when the animal encounters N conditions in extinction it continues to run because these are precisely the conditions under which running has been reinforced. Frustration and the whole r_F apparatus proposed by Amsel (1962) to explain the partial reinforcement effect are no longer necessary; it is only necessary to assume that rats can remember whether they have been rewarded on the previous trial, and that this memory constitutes an internal stimulus which can control the running response. In the previous section it has been noted that there seems to be a natural affinity between the study of memory and cognitive interpretations of behavior. Now it must be noted that there is no necessity to this apparent relationship and that Capaldi's views on these matters present an instructive exception to the rule. Capaldi (1971) appears to hold a rather conservative view of learning; learning is said to involve the reinforcement of S–R associations. The only unusual element in Capaldi's analysis is the powerful assumption that the memory of the previous trial outcome can serve as a stimulus to control behavior.

How long does the memory of the prior trial outcome persist? At least 24 hr, according to several studies (Capaldi, 1967). Evidently the rat is not confused by intertrial feeding; it is the association of reward and a particular place that is remembered. Food given in an entirely different context appears to produce little interference. However, if food is given intermittently in both a black and a white apparatus, then interference effects (intrusion errors) are found (Capaldi & Spivey, 1964). This finding argues for memory contents of the S–S* form, where S is a particular goal box or other food location. It is interesting to note that these memories of whether there has been food, such as Capaldi proposes, persist appreciably longer than memories of where the food has been. Granting that there are hazards in comparing across experiments, Petrinovich and Bolles (1957) found the latter persisting only 4 hr or so. The longer memory for whether is also consistent with results reported by Logan (1960) for rats given one trial a day in an alley.

One other point requires comment, and that is the question of what is learned in Capaldi's experimental situation. Consider that when, say, the rat was fed on the previous trial, there was no time gap between S, the goal box situation, and S*, food. This S–S* association may therefore be formed under conditions of maximal contiguity, and this formed association must simply be stored and retrieved on the next trial if the rat is to respond appropriately. That is, an S–S* view of what is learned requires that appropriate behavior depend on retentive memory, as defined in the introduction; that is, on the retention of a formed association. From the conventional S–R view of learning, however, what is learned must be an S–R association, and for the animal to respond as it does, the association must be between the objective stimulus occurring on one trial and the response that occurs on the next trial. The formation of an S–R association with such a long interval between stimulus and response seems implausible.

Capaldi (1971), who generally endorses the S–R position, resolves the dilemma by making the memory of S the effective stimulus; the memory of S can be contiguous with R and learning under such conditions may well be expected. Capaldi's resolution implies that the content of memory in this case is an isolated event, however, namely, the N or R that has occurred on the previous trial, and this possibility may be less defensible than the possibility that the crucial memory, and that what is learned, has an S–S form. In short, there seems to be no way an S–R approach can account for learning in this situation without adding some kind of memory concept. If what is learned is still assumed to be an S–R association, then we must suppose either that the S term is remembered associatively over an impossibly long period of time or, with Capaldi, that S is simply remembered by itself. A more comfortable position seems to be that what is learned is an S–S association, and that it is remembered retentively from one trial to the next.

There is now good reason to believe that rats can remember how much food they receive. Capaldi and Cogan (1963) found the same kind of alternation in response speed described above when they alternated large and small amounts of reward (AOR). In addition, there is a tradition of conducting AOR studies with one trial a day–Crespi (1942) started it–and although the primary motivational AOR effect can be attributed to a permanent buildup of r_G, the secondary emotional effects, which Crespi called elation and depression, cannot. These emotional effects appear to depend on a discrepancy between expected reward (that is, the memory of past reward) and current AOR conditions. The existence of such effects implies, albeit rather indirectly, the persistence of memory of AOR conditions over a 24-hr period. Gleitman and Steinman (1964) confirmed this speculation in the case of the depression effect by showing that it disappeared over a very long (68-day) retention period. Animals that were shifted down in AOR 68 days after training seemed to have forgotten what the original AOR had been. We are left not knowing just how long memory of AOR persists, but the retention of how much is evidently relatively long compared with the retention of whether and where, discussed previously. It is interesting to note that in Gleitman and Steinman's study the animals had no opportunity to experience different AOR until the time of testing, and they showed prolonged retention. By contrast, when Petrinovich and Bolles made reward contingent on their animals remembering where the previous reward had been, the animals showed much quicker forgetting. Intermediate retention was shown by Capaldi's animals, which did discriminate the different AOR conditions but were not required to do so. We may wonder whether contextual learning, that is, the acquisition of information that is not essential for solving a problem, persists in memory longer than critical information. Perhaps the recurrence of different cues (or different parameters of reward in this case), as is required in discrimination learning, produces interference that destroys the memory of these cues, whereas information about constant stimuli (or constant parameters of reward), if processed at all, is well retained. However, this is apparently not the case, in general. Gleitman (1971) has described research done in his laboratory showing that although irrelevant cues, that is, such contextual stimuli as the color of a runway, are forgotten over retention intervals of 2 months, the significance of discriminative stimuli is not forgotten over comparable intervals. In other words, once a discrimination is learned, the meaning of S+ and S– is very well remembered, whereas the stimulus control of other cues, such as the color of the alley, is lost.

Another line of evidence for memory of AOR is a series of studies of proactive and retroactive interference with changing AOR conditions. The framework for these studies is also provided by the negative contrast effect (depression), originally reported by Crespi (1942). In a typical experiment, one group of animals is trained with a large AOR and is then shifted to a small AOR. This group usually performs more poorly after the shift than a control group trained all along with the small AOR. Suppose, now, that the experimental group is first given some preliminary training with the small AOR. Will this treatment reduce the contrast effect? If it does (Capaldi & Lynch, 1967), then it can be argued that the original experience with the small AOR is remembered when small AOR is encountered again later. At the same time, the experience with large AOR can be expected to produce retroactive interference with the retention of small AOR, which may be demonstrated in a more or less gradual transition, after the shift, from fast to slow running. This kind of paradigm obviously makes possible the systematic investigation of many well-known memory effects using animal subjects. Spear (1967) has reviewed a number of such studies, many conducted in his own laboratory. The evidence makes a compelling case, first for the idea that animals have good memories for AOR conditions, and second for the use of this paradigm to study animal memory.

Other dimensions of reward are undoubtedly remembered too, but the direct evidence for the memory of other dimensions is scant. Pschirrer (1972) has shown that the kind of reward (milk vs rat pellets) received on one trial can be remembered and used as a discriminative cue for responding on the next trial. Such memories have only been shown to persist for 15 min, but it may be noted that Pschirrer's animals were run in a discrimination problem where the constant changing of reward conditions might be expected to produce interference. There is also indirect evidence that rats can process and retain information about the temporal distribution of reward. Herrnstein (1970) has emphasized that in many operant conditioning situations an animal's rate of responding appears to depend not so much on an instantaneous response-strengthening effect of discrete rewards as on the number of rewards received over a period of time. It is not that each reward produces an increment in response strength but that the rate of reward over time controls the rate of responding. This phenomenon is most clearly seen in choice situations where one response is rewarded, say, twice as frequently as a second response. According to Herrnstein, the animal will tend to match its relative rates of responding to the relative rates of reward, in this case, in a 2:1 ratio. If this occurs, we may think of the animal remembering, and using the information about, when reward has occurred. Moreover, whereas the existence of scalloping with FI schedules suggests that animals can remember when the last reward was, the matching law suggests they can remember when the last several rewards were. It seems necessary that they remember the timing of an appreciable number of past rewards if they are to match accurately. This phenomenon does not seem to have been used explicitely for the study of memory, and appropriate controls have not been run to demonstrate that it really is a memory phenomenon. A further difficulty here is that although the matching phenomenon appears to call the conventional S–R view of learning into question, S–R theorists have met ("anticipated" is more correct) the challenge. Some years ago, Estes and Straughan (1954) showed how a simple incremental learning model could account for matching.[3] Still, Herrnstein's analysis reflects a changing perspective among reinforcement theorists of how reinforcement works and of what learning via reinforcement consists. The student of animal memory may well discover here a new kind of fruitful relationship between learning and memory.

This section has been concerned with the memory of reward and with specific memories of the dimensions of reward, such as where, when, and whether it occurs; what kind of reward it is; and how much of it there is. This emphasis reflects, in part, my bias about what I think animals learn. I believe that the largest part of animal learning consists in assimilating and storing information about S–S* relationships in the environment. Because S*s are important, it is important that animals be able to learn about them, what they are, and where and when they occur. I believe that, in general, animals behave adaptively when they are able to anticipate S*s and that they can do this most effectively by remembering the S–S* relationships that characterize their worlds.

THE GARCIA EFFECT

So far, this chapter has dealt with retentive memory, that is, the retention of a learned association over a period of time. Retentive memory phenomena have never really challenged the learning theorist. Indeed, the learning theorist takes the persistence of learned associations for granted; he tends to be surprised at instances of forgetting. He has always held a quite different view of associative

[3] More recent research, for example, Rose and Vitz (1966), has shown that human subjects do learn something about particular sequences and runs of events. So Estes' early statistical learning model, which assumes that there are no sequential effects, serves only as a first approximation to the subjects' behavior. Later mathematical models have been elaborated to better approximate the facts; and some of these newer models (e.g., Restle, 1961) make explicit assumptions about memory.

memory, however. For example, Spence (1947) maintained that if reward were delayed more than just a few seconds no learning could occur. According to Hull (1943), an asychrony between stimulus and response of more than a few seconds would prevent learning. In short, whereas retentive memory was perfect, associative memory was virtually nonexistent. However, John Garcia discovered a phenomenon that must cause all students of animal behavior, whether they are concerned with learning or with memory, to stop and take a careful look at some of their most cherished convictions.[4]

When a rat drinks a novel substance, such as saccharin, and is then made sick by an injection of apomorphine, it is subsequently likely to show an aversion to saccharin. It is likely to drink very little saccharin if offered it the next day. This basic phenomenon (reviewed by Revusky & Garcia, 1970) is called a "conditioned food aversion"; it is called "conditioned" because it is evidently based on learning, and it is called a "food aversion" because it is primarily the acceptability of the specific substance, saccharin, that is affected. Control animals made sick but not given saccharin first, or control animals given saccharin but not made sick, will drink saccharin when offered it the next day. It therefore appears that the experimental animals' aversion depends specifically on the association of saccharin with sickness. This is the basic phenomenon that Garcia discovered.

The Garcia effect has considerable generality across experimental conditions. It has now been reported in a number of animal species (but there are few, if any, convincing demonstrations in human subjects); it can be produced by a great variety of toxic agents; and aversions can be produced to a wide range of foods and solutions. However, there are some important limitations to this last claim. Revusky and Bedarf (1967) showed that just a little prior familiarity with the test substance would sharply reduce the conditioned aversion to it. Substantial aversions are only found with novel substances. In addition, Weisinger, Parker, and Skorupski (1974) have found that certain combinations of test substance and toxic agents fail to yield aversions. Altogether, however, the Garcia effect is a robust phenomena with considerable generality. Even so, there are some further specificities that I want to emphasize.

One of the most interesting and theoretically important specificities of the Garcia effect is the specificity of stimulus factors that enter into an aversion. For example, Garcia and Koelling (1966) found that although rats would avoid the distinctive taste of saccharin that was associated with illness, they would not avoid plain water that was made distinctive by flashing a light and sounding a

[4] Some of the results noted in the previous section, such as the patterning of running speed in conformity with the pattern of reward, can be viewed as associative memory effects. However, these results are also subject to alternative interpretations, as we have seen. An associative memory interpretation of the Garcia effect is not so easily dismissed. It should be clear that here I am following Revusky's (1971) analysis of the Garcia effect. It is a thoughtful and thought provoking account, and it is difficult not to follow it, but I will part company with Revusky shortly.

clicker whenever they drank it. That is, their rats would learn to avoid a substance with a distinctive taste but not a substance with distinctive audiovisual qualities. By contrast, Wilcoxon, Dragoin, and Kral (1971) found that Japanese quail would avoid the visual appearance of a test food associated with illness but not a test food with a distinctive taste. These results suggest that each species of animal has its own hierarchy of cues that enter into food aversions; for some animals taste is high in the hierarchy, for others visual cues are higher.

Let us pursue this matter further. Consider the procedure used by Wilcoxon and co-workers. They gave a group of quail blue, salty water to drink and then made them sick. On the next day, after the birds had recovered from the illness, the group was divided in two, and half were tested with blue, unsalty water. They drank little of it, less than control birds offered blue, unsalty water for the first time. Therefore we may safely conclude that for the quail the blue color provided a highly salient cue for aversion learning. The other half of the birds were offered uncolored, salty water in the posttest, and they showed no aversion to it, relative to controls. Now this negative result is subject to several possible interpretations. One possibility is that at the time of the original exposure the birds do not perceive the unusual taste (many birds have a poor sense of taste). Another, more interesting possibility is that the salty taste is perceived at the time, but the birds pay little attention to it because it is already relatively familiar to them. There might have been appropriate sensory input, therefore, but this information was immediately discounted instead of stored, so that it could not be remembered at the time of the posttest. The most interesting possibility (to me) is that the salty taste was perceived and remembered, but this information was discounted and not used at the time of the posttest. We do not yet know which possibility will ultimately prove to be the most appropriate interpretation of these particular results. However, we do know that each of these mechanisms can operate in other, related situations.

Consider the possibility that information is input and stored but not used. Something like this seems to have happened in a study, not yet published, by Anthony Riley at the University of Washington. Riley was trying to get rats to avoid a particular place where drinking was followed by poisoning. The rats were given their daily water ration in a fixed location. After several days of this, the water bottle was put in a new location, and after the drinking session that day the rats were poisoned. Then on the following day the rats were given two water bottles, one in the old "safe" location and one in the new "poison" location. The consumption data indicated a strong aversion to the poison location. These results therefore show that the rat can learn to avoid a poison-related place, as well as a poison-related taste. There is nothing too surprising about that. The most interesting results were those Riley found with another group of animals treated in exactly the same way but tested with water in one bottle and a novel saccharin solution in the other. Given the choice between drinking water and saccharin, they showed a strong aversion to saccharin. It made no difference whether the saccharin was in the safe place or the poison place. Notice that if we only had the data from this last group we would have to conclude that the rat learns nothing about the place of poison, because the aversion to saccharin was independent of its location. However, the data from the first group show that information about the poison location has been stored and can be retrieved if there is water at both locations (or, as it turns out, saccharin at both locations). It is as if the rat attributes its misfortune to location cues if that is all the information it has, but in the presence of taste cues it simply fails to utilize this information and instead attributes its previous troubles, erroneously, to the taste cues. It is hard to see how any conventional S–R account of learning can explain such a thing.

The Garcia effect poses other problems for the learning theorist. One serious problem is that food aversions are learned even when the consequence, illness, follows some hours after the antecedent, ingestion of the test substance.[5] It has been recognized since the time of Aristotle that contiguity plays an important part in association formation. Because contiguous events are readily associated, it has been easy to generalize the conclusion that contiguity is necessary for learning. Contiguity permits the "trace" of the antecedent to be still present when the consequence occurs, and it is easy to go on from there to Spence's (1947) conclusion that there can be no learning if the consequence is appreciably delayed (delayed reinforcement can only be effective if immediate secondary reinforcement is interposed). However, this conception of learning is torn assunder by the discovery that an association can be formed with a long-delayed consequence. We can see at once that contiguity is not a necessary condition for learning. We can also see that food aversion learning must be a memory phenomenon. When the rat becomes ill, it must remember specifically the taste of the food it ate hours before. Here is the interesting question, however: Why does the rat remember specifically the taste of the novel food when it becomes ill later? Why does it retrieve this specific event to associate with illness, instead of any of the intervening events? The psychologist can see that the earlier ingestion is relevant to the subsequent illness, but the question is why does the rat see it as relevant? It does, because the association is rapidly formed and vividly expressed as a food aversion, but how does this happen?

Revusky (1971) has suggested that this learning depends on ordinary associative mechanisms but that it enjoys the special advantage of being free from interference. During the delay interval the rat experiences no other taste stimuli and, presumably, no other major changes in bodily welfare. Other intervening events are different enough not to cause much interference. The argument seems

[5] There are several reasons for thinking of aversion learning as an S–S* association between taste and illness, rather than as an instance of S–R learning where illness weakens a connection between taste and the consummatory response. One reason for this interpretation is the finding by Domjan and Wilson (1972a) that aversions can be learned by rats that are curarized so that they cannot make the consummatory response. When saccharin was trickled over their tongues and the rats were then made sick, they later showed an aversion to saccharin.

to make sense when applied to the rat and aversions to taste cues. However, it does not seem appropriate for the bird and its aversion based on the sight of food. Why does the multitude of intervening visual stimuli not produce massive interference? Revusky's answer is that there is an additional mechanism, which he aptly calls "stimulus relevance." Apparently, because irrelevant stimulus events are irrelevant, they produce no interference. I think that as strange as this concept sounds the idea of stimulus relevance is valid and valuable, granted that it must be articulated more clearly. My main complaint with Revusky's analysis is that he does not carry the stimulus relevance principle far enough. It seems to me that if stimulus relevance can "protect" the memory of the test substance, then there is no need for a freedom from interference principle to protect it. Moreover, the stimulus relevance principle can explain not only the persistence of the memory for the cue, but its association with the consequence. The interference argument cannot do this; it explains why the cue memory is kept in storage, but not why it is retrieved at the critical time when S*, illness, occurs. It is clearly not sufficient that some stimulus property of the dangerous food be remembered; this information must also be retrieved when the animal becomes sick. Moreover, the illness must be associated with this remembered food cue and not with all the other things the rat may have stored in memory. How does this superb relevance principle work—that is the great puzzle.

One possibility is that whenever the rat eats a novel food, the taste and other properties of the food are stored in a special-purpose buffer, a sort of intermediate-term memory store, and this information is processed only when there is subsequently a new state of illness or wellbeing. Other S*s seem to have little effect on the acceptability of new foods, and other kinds of antecedent inputs do not seem to get associated with illness (Garcia & Koelling, 1966). So it appears that stimulus events that enter into the Garcia effect are stored in a relatively closed system. The special memory system, or portion of the general memory, that makes such learning possible appears to serve this function rather exclusively. The proposal, then, is that such animals as the rat have evolved a specialized memory system, or function, which stores information about ingested materials and lets this information be retrieved when there is a change in the digestive system. It is hard to see how else the antecedent information can be stored and then associated with another, long-delayed event. It also seems likely that the temporal properties of this memory system, which are described in the next paragraph, have evolved to match the typical time course of the harmful and/or beneficial effects that follow eating. Other parts of the total memory system may be regarded in the same way. For example, if painful events in the natural environment are often closely contiguous with danger signals, then we should not be surprised to find that contiguity is an important parameter in fear conditioning—which it seems to be. An animal's psychological capacities have to reflect the causal texture of the environment (as Tolman once noted), if it is to cope adaptively with its environment.

There is also, apparently, some kind of forgetting of the food cue. Typically, very strong aversions are found if illness is delayed $\frac{1}{2}$ hr, modest aversions are found with 4-hr delays, and little aversion is found with delays much longer than 4 hr. The "delay of reinforcement" gradient is therefore about 1000 times longer than that usually reported in instrumental learning situations. The remarkable length of the gradient has already been discussed; now we must inquire into why there is a gradient at all. That is, why does the animal ultimately forget the taste of a novel food? Forgetting may be attributed to interference or to a decay process, but there are facts that argue against either of these versions of a forgetting model. One difficulty with a forgetting interpretation is that any prior exposure to the test food greatly reduces the size of the aversion that can later be conditioned to it (Revusky & Bedarf, 1967). It is apparent then that the properties of the test food are not forgotten. With the passage of time the test food does not become novel again; it is simply regarded differently. Kalat and Rozin (1973) have helped us pin down what this difference is. Their argument is that over a long delay between ingestion and illness there is not a passive forgetting but an active learning going on. What the animal learns is that the novel food is safe. Safety is learned slowly, they suppose, so that if illness occurs shortly after ingestion, little safety will be learned and there will be little interference with aversion learning. If illness is delayed for, say, 4 hr, however, the animal will have had appreciable time to learn that the food is safe (actually, it will have been safe for 4 hr). The stimulus properties of the food will become associated with a different S*, safety, and this learning will interfere with the learning of an aversion to the food. Kalat and Rozin (1973) tested their learned-safety hypothesis against the forgetting hypothesis in the following way. They gave a group of rats the test food twice, once 4 hr before poisoning and again ½ hr before poisoning. These animals showed only a modest aversion, comparable to that of a control group just poisoned 4 hr after ingestion (and which had, therefore, a comparable amount of time to learn about safety), and much less of an aversion than a control group poisoned just ½ hr after ingestion (which, therefore, had the same amount of time for the "trace" to fade). Kalat and Rozin's results might have been caused by their use of a high-calorie test food. That is, the rat may learn that a food is good for it, in addition to being safe. However, the same results have been obtained by Bolles, Riley, and Laskowski (1973) using nonnutritive saccharin as the test food. Recent results by Domjan and Bowman (1974) and Nachman and Jones (1974) indicate that the learned-safety mechanism cannot provide a complete account of the S–S* temporal gradient; there is also, apparently, some sort of forgetting of S. However, their results do not require us to abandon the learned-safety idea either.

This learned-safety mechanism appears strange indeed from the viewpoint of traditional S–R theory reinforcement theory. It is learning, because it interferes with subsequent learning, but where is the reinforcer? Where is the response?

Whereas one might hope to find a response characteristic of illness, or a response that might be identified as an illness reaction, where would one look for a safety reaction that competes with it? From a more casual, broadly functional viewpoint, however, it all makes sense. The biological function of the Garcia effect is to protect the animal from noxious foods. Consider an animal encountering a new substance for the first time. It may be toxic or it may be beneficial. The optimum strategy, and Barnett (1958) has indicated that it is in fact the wild rat's strategy, is to eat a little and then wait and see. If nothing happens in the next few hours, then the food is judged to be safe. If instead the animal becomes sick, then the food is judged to be bad. The temporal gradient of safety learning and/or food aversion learning is too much like the typical time course of digestion and food poisoning to be just a coincidence. The system appears to have evolved expressly to serve this function. What it looks like psychologically is what Krechevsky (1932) once called "hypotheses" in rats. The animal has an initial hypothesis: this novel food is bad. If the animal eats it and then soon becomes ill the hypothesis is confirmed and the food is subsequently avoided. If, however, as time passes, the animal is all right, the initial hypothesis is disconfirmed, rejected, and replaced by the useful new hypothesis: this food is good.

This concept of an initial hypothesis, a sort of innate suspiciousness, leads us to an interesting complication of the Garcia effect that must be discussed before we can proceed. Barnett (1958) has emphasized that wild rats show a strong neophobia reaction to new things, particular to new foods. More recently, Mitchell, Scott, and Williams (1973) have found comparable but somewhat weaker neophobic reactions in laboratory rats. Subsequently, it has been demonstrated (Riley, 1974) that the initial neophobia to novel foods in laboratory animals is enhanced by recent illness. An animal made sick one day drinks less of a novel saccharin solution the next day than a control animal that has not been sick. Moreover, both animals drink less saccharin, on the average, than other animals that have consumed saccharin before. There appears to be a normal neophobia in normal animals, therefore, and an enhanced neophobia in animals that have been sick. Now this enhanced neophobia can be regarded as a kind of pseudo-conditioned food aversion; it is a reaction to an S, as a result of S*, without S and S* having been paired. It is clear that because this pseudo-conditioned aversion mimics a learned aversion, research on the associative processes involved in the Garcia effect should include controls to assess the contribution of the neophobia effect. When we do this in our laboratory (Riley, 1974), we find that the aversion attributable to neophobia is, very roughly, about one-half of the total aversion. It therefore appears that there are two mechanisms that help the rat avoid potentially dangerous foods. One is that when the rat has recently been ill, it has a general aversion, that is, a nonassociative or unlearned aversion, to novel substances. The second mechanism is that the rat has a specific aversion, an associative or learned aversion, to novel substances ingested just before becoming sick. Having split these mechanisms apart, in the next section we will see whether we can put them back together again in a unified manner. To conclude this section, we may merely observe that the Garcia effect has a number of far-reaching implications for the study of both learning and memory in animals. It has already become apparent (Bolles, 1975a) that many psychologists in the learning area have been inspired by Garcia's research to look for new concepts and new paradigms to understand learning. We now see more clearly that the laws of learning must be placed within a broad biological context. Although the whole tradition of animal experimentation tells us that long-delayed learning with a single trial is impossible, rats must be capable of food aversion learning–or there would be no rats. So we must ask what we have been doing wrong during the tradition. We can also now see more clearly the diversity and specificity of learning mechanisms; we can no longer believe in a monolithic S–R model. The student of animal memory can also benefit from these new discoveries; he has a number of new associative memory phenomena to reckon with; and he has a simple but powerful experimental procedure with which he can pursue his study of memory.

THE ATTRIBUTION CONCEPT

There is one sense in which a memory interpretation provides an incomplete account of behavior. By itself memory does not produce behavior; some additional principles are necessary to carry memory over into a response. There is no problem if the contents of memory are assumed to be S–R connections, because then, in effect, it is the response itself that is remembered or forgotten. However, what if memory consists of isolated stimulus events or of S–S* representations?

Capaldi (1971) has offered one solution to the problem, as we have already observed. He treats the memories of stimulus events as stimuli, something like Hume's "faint impressions" of earlier sensations. Then these new stimuli are introduced into the familiar S–R model. Elsewhere (Bolles, 1975a), I have suggested an alternative solution: when an animal encounters a predictive cue, that is, an S that arouses an S–S* expectancy, it responds in relatively fixed, species-specific ways directly to the S. Examples are the rat freezing in a box where it expects shock and the pigeon pecking a key in a situation where a light on the key is correlated with food. Other solutions to the response problem have been suggested by other writers.

There is a further problem that must be solved before we can have a complete account of behavior. This is the difficult and perhaps obscure problem of how heterogeneous associations are to be conceived. By "heterogenous associations" I mean associations between dissimilar and disparate stimulus events. Black and white colors are easy to associate and it is easy to conceive of such an association, because they share a common stimulus dimension. The basis of the association is the homogeneity of the elements. The same is probably true, in some sense, of the verbal materials used with human subjects. How are disparate elements associated, however? Contiguity helps, no doubt, but can we be sure that contiguity alone makes the rat learn that this box means food and that this tone means shock, or may some other principle be involved? The rat appears to be peculiarly capable of learning about places; the rat is a natural geographer. So even though there may be relatively little contiguity between, say, a particular place and the presentation of food, and even though these events are heterogeneous, the rat readily associates place and food. The basis for this association seems to be partly some characteristic of the rat, something in how its associative processes are organized. The Garcia effect provides a striking illustration of the problem I am trying to articulate. Some birds have an awesome capacity, as we have seen, to associate the visual appearance of a novel food with illness. What is the basis of this association? It cannot be contiguity, because appreciable delays do not prevent it, and it cannot be homogeneity because the stimulus events are in no sense similar. How is such a heterogeneous association formed?

One possibility is suggested by a provocative paper by Testa (1974). The suggestion is that disparate events may be associated when they are perceived as being causally related. Testa reduces the causality relationship, as Hume did, to temporal sequence and similarity of temporal pattern, but it can also be accepted as is, without the reduction. Testa applied this analysis to the explanation of some avoidance-learning phenomena; here the idea of causation is applied to the explanation of heterogeneous association. To put it loosely, perhaps when the rat becomes ill it attributes, in some sense, its illness to what it has eaten. Perhaps the food is perceived as causing the illness. People make such attributions explicity, and maybe the rat does the same thing implicity.

The notion that people attribute causation in characteristic ways was introduced into the psychology of human motivation by Heider (1958). He noted, for example, that one person may characteristically attribute the good things that happen to him to his own virtues and the bad things that happen to him to bad luck. Such tendencies tell us much about the individual and give us considerable insight into human motivation generally (Weiner, 1972). It is doubtful that the rat has the same widespread and personality-determining attribution tendencies as the human, but it seems worthwhile to look for a comparable mechanism. Moreover, it seems important to translate such novel concepts into experimental procedures.

Let me describe an attribution study recently carried out in my laboratory by Kam Major. This was a neophobia experiment; one group of rats received nothing novel, just their accustomed daily water ration, before being injected with poison (LiCl). When this group was tested the next day with a novel saccharin solution, they showed a neophobic reaction to it. Normal rats prefer saccharin about two to one over plain water, but these animals chose water over saccharin three to one in a two-bottle test. Our hypothesis was that neophobia in this kind of situation results from the rats not being able to attribute illness to anything. They avoid all novel substances because they do not know to what to attribute their illness. For the experimental group the situation was structured differently. These animals were given a novel saline solution before they were poisoned, so that their illness could be attributed to the saline. On the two-bottle test the next day these animals should show less aversion to saccharin. That was the prediction from the attribution hypothesis. Something went wrong: There was no difference between groups. Both groups showed the same strong aversion on the first test and the same course of habituation of neophobia (3 days) on successive tests.

When Wallace Wilkins was at the University of Washington he did a small attribution study using fear conditioning. The control animals were hungry rats that were put into a novel box with food, but before they started to eat they were given a severe shock. The shock was sufficiently intense to delay the onset of eating about 1 hr; it presumably took that long for fear of the apparatus to dissipate. Wilkins conjectured that fear was so slow to dissipate in these animals because they had nothing tangible to which to attribute it. For the experimental group everything was the same except that a distinctive wooden object was put in with the rat, shock was administered when the rat approached it, and then the object was taken out. With a discrete, tangible "thing" to attribute their troubles to, these rats should have had less fear of the situation and should have commenced eating sooner than the control group. They should have, but they did not. There was no difference between the groups.

The reader may decide for himself whether (1) we did not conduct the appropriate experiments, (2) we did not understand how rats attribute causality, (3) rats do not attribute causality, or (4) the whole enterprise is best forgotten.

INSTRUCTIONS AND EPISODES

One of the great conveniences of using human subjects is the ease with which they can be given instructions. To be sure, they often follow their own ways instead of following the instructions they are given. Human subjects often bring to an experimental task their own ideas about how to interpret the task and their own tricks for working on it. By and large, however, they are remarkably compliant with experimental instructions. Their versatility is exploited; because humans have a history of following varied instructions it is easy to get them to do new things for us. Because there is an arbitrariness to the tasks they can be put on, we can put our human subjects directly onto those tasks that clarify what we want to investigate. We can adapt them to the study of our paradigmic puzzles, make them fit our mold.

Animals are quite different, however. The conveying of instructions to animals can be painstaking and uncertain business, compared with which the final phase of the experiment is simply and quickly carried out (Gleitman, 1974). Animals may ultimately be induced to do a variety of remarkable and unlikely things but it may take enormous ingenuity and patience to convey the appropriate instructions. Roger Davis' work, described in chapter 7, illustrates that primates can, indeed, be trained to perform on subtle, complex, and unlikely tasks—but look at the prior training his animals had received! In contrast with sophisticated human subjects, naive animals have very strong proclivities to follow their own ways and to ignore the instructions we give them. They tend to have very set ways of perceiving the task at hand and rather species-specific tricks for solving what they perceive to be their problems. So instead of versatility we find constraints on what the animal can do. Rather than assign them arbitrary tasks we have to put our animals on tasks that conform to their ways of doing things; we have to adapt ourselves and our experimental procedures to them.

For a long time few psychologists accepted the fact that there are constraints in animal learning. One reason was that such an idea compromised the importance and universality of the learning principles that had been developed. Today, however, increasingly, we recognize that there are constraints on learning and limitations to the established principles of learning (Bolles, 1975b). We now recognize that there is a "structure" to learning in animals. What an animal learns is only in part a reflection of the environmental contingencies it has been subjected to, it also depends on the animal. Learning probably depends as much on the animal's evolutionary history as on its personal reinforcement history. The great benefit that derives from this broader view of animal behavior is that in studying the learning of some animals we may hope to discover something important about the animal. With the old philosophy, the study of learning could only tell us about learning.

If learning has an inherent structure, which is relatively species specific, we may suppose that memory does too. Tulving (1972) has proposed that in many human memory studies the subjects' principal task is to sort out from all of the familiar, well-known materials stored in memory those which occurred during the experimental episode. The remarkable thing is that subjects can do this at all, that is, that there is any episodic memory. How can a subject learn to give antonyms to one list and synonyms to another, and keep straight what he is doing? He does it, and without distortion of the permanent semantic memory, by treating the items as elements of the episode. The versatility of human memory is shown by the variety of materials that can be used to define an episode, and by the arbitrary nature of functional episodes. We may expect animal memory to be much more restricted. We may expect that only certain classes of events can be put together to define an episode, and that only certain kinds of episodes can be recognized as such. Whereas the human subject can play all sorts of different games, the animal takes his S*s too seriously to recognize that they may enter into different episodes.

Perhaps oddity problems can be regarded as episodes. Much of the time, however, episodes are rather basic and primitive, at least for the rat. Food in a particular place—that is the sort of category into which the rat's memory is organized. The typical episode for the rat consists of a variety of Ss and an S*. This view implies that to impose an arbitrary learning task on an animal is to invite a failure of learning and/or memory. The corollary is that with an arrangement of the experimental situation that is appropriate—appropriate for the animal—prodigious feats of learning and/or memory may be observed. The Garcia effect is a good example of what I mean.

Another example is the relatively long-term associative memory phenomenon reported by Lett (1973, 1975). In the first study Lett trained rats to go to one side of a T maze, but instead of feeding an animal for making a correct response, she removed it from the apparatus, waited several minutes, and then returned the animal to the start box and gave it food there. In other words, a correct response was rewarded at the beginning of the next trial rather than at the execution of the response, and reward was given in the start box instead of the goal box. This procedure is ridiculous from the perspective of how we have come to think learning experiments should be conducted, but the rat understands it. Everything was done wrong, it seems, but the rat solved the problem, and the long delay of reward proved to be little handicap. The secret of Lett's procedure is evidently removing the animal immediately after the response, so as to preserve the contextual cue value of the apparatus. A follow-up study (Lett, 1975), using the same procedure to establish a black–white discrimination, shows that leaving the rat in the empty goal box for just a few seconds prevents learning. We may think of the apparatus defining for the rat a natural kind of episode, and with this well-defined episode the lapse of real time is irrelevant.

In closing, I would like to describe an unpublished experiment by Michael Wilson and me. The results indicate that forgetting cannot always be taken at face value, that a failure of association may be caused by not properly defining the experimental episode rather than by a failure of memory, and that rats can remember and/or learn better than one may suppose if they are given the proper instructions. The setting for our study was the famous experiment by Grice (1948), which has been widely cited as demonstrating that animals have virtually no associative memory. Grice trained rats on a black–white discrimination; on passing through the black or the white compartment, the rat could be detained in a gray compartment and then permitted to move on to a gray goal box. The gray areas of the apparatus precluded the mediation of discrimination by secondary reinforcement, for whatever secondary reinforcement value the goal box and its gray color might have acquired, there should have been no differential reinforcement (or incentive motivation) associated with the black and white cues. The only basis for learning the discrimination, Grice argued (and I concur), would be for the rat to remember the color of the cue while detained in a gray compartment. Grice found rapid learning with zero delay, but virtually no learning with delays of 5 sec or more. He (and Spence, 1947) concluded that the rat's association memory was limited to a second or two, and that any learning found with longer delayed reinforcement must be attributed to secondary reinforcement.

It is clear that something prevented learning under delay conditions, but was it forgetting? There are other possible mechanisms. Lawrence and Hommel (1961) suggested that in this situation rats might be confused about how many goal boxes there were. If the boxes were not discriminated, then the animals would have the unsolvable problem of why there was sometimes food and sometimes not. They found that making the boxes discriminably different alleviated the delay decrement. This finding discredits the memory interpretation, but it presents a new puzzle: why is the rat able to learn two complicated associations over a delay (i.e., white cue means one goal box and food, and black cue means another goal box and no food) but neither of the simpler associations Grice required (i.e., white means food, or black means no food)? Why should the "instructions" about there being two goal boxes make the problem solvable? Do associations involving location mediate other associations for the rat?

A second mechanism that may make the problem unsolvable is confusion about its solvability. When Grice's animals failed to learn they displayed position habits. It occurred to Wilson and me that reward on 50% of the trials might maintain a position habit, and that a cure for this difficulty might be to give animals instructions about 100% reward. We replicated Grice's original study as closely as we could with a control group and we also found no learning with a 10-sec delay. However, we found that an experimental group with a 10-sec delay previously given 40 forced trials to the white cue learned the discrimination in 52 trials. It appears that a generalized expectancy about the solvability of the problem, that is, about the probability of reward, makes the problem solvable. This finding, too, discredits the loss of memory interpretation, but there is another puzzle. We ran another control group; it had the same 10-sec delay and it received the same pattern of 40 forced preliminary trials. The difference was that during the forced preliminary trials the entire apparatus was gray. Therefore, this group was given the same instructions about the solvability of the problem but was told nothing about the positive cue. This group failed to learn. It was not just that for the experimental group responding to the positive cue had been rewarded an additional number of times. We ran one more group that received a number of preliminary forced trials to the positive cue but was rewarded on only half of these trials, and this group also failed to learn. Therefore, it appeared that to be effective our instructions had to tell the animal something about the positive cue as well as something about the solvability of the problem. Perhaps the generalized expectancy of 100% reward must be tied to the cue itself.

There are other aspects of this situation I do not understand. One is that there still remains a sense in which the Grice–Spence interpretation is valid. No matter how confused the rat may be about how many boxes there are, or how often food is down there, it learns the discrimination when there is no delay. Contiguity appears to be a powerful basis for association, even if it is not the sole basis. Indeed, it may only be by eliminating the powerful force of contiguity that we will be able to discover how the memory of an animal is organized. Like most problems in animal learning and animal memory, this one will yield only to further study.

CONCLUSION

Learning and memory are interrelated areas. Even though the early learning theorists minimized the importance of memory, they made substantial but indirect contributions both to the theory of memory and to the discovery of memory phenomena. Now the tables are beginning to turn. The animal memory area has developed to the point where it presents phenomena the learning theorist cannot ignore, and it offers new, powerful theoretical tools, which the learning area can well use.

Although memory concepts are amenable to an S–R analysis, they seem to have a more natural affinity with a cognitive analysis of behavior, that is, one that emphasizes information processing and asks questions such as what information gets conveyed from environment to organism, and what information is stored and later retrieved so that the organism can respond adaptively to its environment. However, our attraction to a cognitive approach and our use of cognitive language should be tempered with an appreciation for the limitations of our animal's cognitive abilities. Just as we are only now becoming aware that learning in animals (and man) is limited, constrained by native predispositions, and selective, in general, so we may expect the kinds of information an animal can input, store, and retrieve to prove limited also. Memory too must be selective, and our task consists largely in discovering its structure.

An important question regarding the structure of memory is whether animals remember isolated events or relationships between events. Does a CS act like a retrieval cue to arouse the memory of the US? Or is the whole specific relationship between CS and US remembered? There is a little evidence suggesting that the rat can readily learn and remember different kinds of S–S* relationships. The rat is apparently adept at processing and retaining information about food rewards—where, when, how often, and how much food it has received.

The study of animal memory poses functional as well as structural questions. What is the basis of association; what makes two events become associated?

Contiguity alone is often sufficient. However, how does an animal learn about the relationship between a cue and a consequence when these events are not contiguous? For example, how does the rat come to associate the taste of a novel food with a subsequent illness? The realization that there must be bases for association other than contiguity opens up a variety of intriguing theoretical possibilities. One possibility is that animals associate events when they can attribute a causal relationship between them. Another possibility is that certain events may be encoded into relatively closed systems, and that they become associated simply by virtue of being in such a special system. Such a system is like an episode, a naturally occurring episode, in which events are associated and/or remembered simply because they constitute the episode. In such a closed system the lapse of time may become almost irrelevant; the rule of contiguity may be defeated. Food-aversion learning is a possible example. Can episodes of an arbitrary sort be induced experimentally? Can we, in effect, "instruct" animals that certain events are related? Such questions may lead us to a better understanding of the structure of learning, the structure of memory, and a deeper appreciation of the behavior of animals.

ACKNOWLEDGMENTS

Supported by National Science Foundation Grant GB-40314.

3

Delay of Consequences and Short-Term Memory in Monkeys

M. R. D'Amato
Jacquilyn K. Cox

Rutgers University

ANIMAL SHORT-TERM MEMORY

As a research topic, animal memory has enjoyed a long history in experimental psychology, witnessed by the fact that almost half a century ago Schneck and Warden (1929) were able to marshall 30 studies that concerned, at least in part, retention in animals. Indeed, many of the concepts that gained currency in human memory research during the subsequent two decades were borrowed from the animal literature. For a variety of reasons, however, research in animal memory failed to keep pace with work in human memory and by the time that interest in the topic had been reawakened, research in human memory had produced and greatly elaborated such concepts as the STM–LTM distinction and the notion of STM as a limited-capacity rehearsal-storage mechanism. Everyone can agree that these concepts have been very useful in rekindling research interest in animal memory, particularly STM. Nevertheless, viewed from the vantage point of an accumulating data base, it now appears that they have a limited relevance for animal STM.

We became skeptical about the utility of the STM–LTM dichotomy and the parallel limited-capacity storage mechanism after we studied delayed matching to sample (DMTS) behavior in *Cebus* monkeys in some detail (D'Amato, 1973). Several things did not square with human models of STM based on these conceptions. First, based on their DMTS performance our animals showed a remarkable STM capacity, in the order of minutes. This was in sharp contrast to the 30–60-sec limit incorporated into a number of human STM models (cf. Norman, 1970). Furthermore, the STM capacity of our animals as measured by DMTS was not a static quantity, as one might expect on the assumption that it mirrored a basic structural property. Instead, it increased, seemingly without end, with accumulated practice. Animals that could scarcely match above chance with an 18-sec retention interval ultimately, after years of practice, became capable of high performance levels with retention intervals as long as 2 or 3 min. Figure 3.1 demonstrates both of these points for one of our more talented subjects, Roscoe, that served in a number of DMTS experiments over a period of some 6 years. During this time, Roscoe was transformed from an animal able to perform only slightly better than chance at a 9-sec retention interval to one that displayed substantial retention with a 9-min delay, an increase of 60-fold. Substantial retention in DMTS tasks with delay intervals of two or more minutes has also been reported for rhesus monkeys (Mello, 1971) and a dolphin (Herman & Gordon, 1974). Pigeons, in contrast, are far less proficient at DMTS, often displaying little evidence of retention with delay intervals of only 10 or 15 sec (e.g., Grant & Roberts, 1973; Roberts, 1972b).

These and other facts led us to the view that the kind of memory studied in DMTS and related animal retention tasks was more likely based on temporal discrimination processes than on the limited-capacity storage mechanisms postulated for human STM. The argument in brief is that when an animal is confronted with the sample stimulus and one or more comparison stimuli at the end of the retention interval, its task is to decide which of the various stimuli it has seen most recently. Essentially this amounts to forming a temporal discrimination between the "time to last seen as sample" for each of the choice stimuli. As a concrete illustration, suppose that a square has served as the sample on the previous trial and after a 1-min intertrial interval, the current trial, on which the sample is a triangle, begins. If the current and previous retention intervals are both 2 min and the choice stimuli on the current trial are square and triangle, the "time to last seen as sample" is 5 min for the square and 2 min for the sample, 2.5 expressed as a ratio. With a 10-sec retention interval on the current trial, this same ratio increases to 19.0. Of course, the actual situation is likely to be far more complex than is suggested by the present illustration, but this general approach can, in a qualitative way, account for both retention gradients and the fact that, within limits, performance in animal and human short-term memory tasks increases with longer intertrial intervals (Herman, 1975; Jarrard & Moise, 1971; Loess & Waugh, 1967). It also accounts for the fact that DMTS performance generally improves as the number of stimuli in the sample set increases (Herman & Gordon, 1974; Herman, 1975; Mason & Wilson, 1974; Worsham, 1975), and it provides an explanation for certain trial sequence effects in DMTS (Herman, 1975; Worsham, 1975). There is, indeed, sufficient qualitative evidence in support of a crucial role for temporal discrimination factors in animal STM to warrant the development of more formal models that take into account not only the sample as a controlling stimulus but other factors as well,

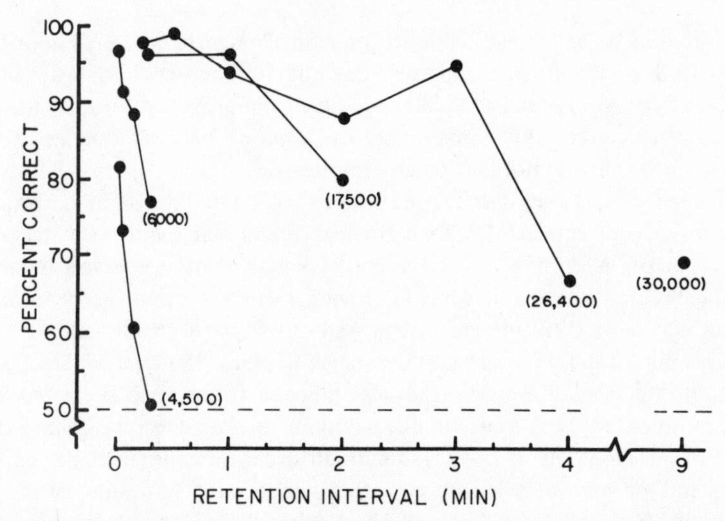

FIG. 3.1 The development of visual short-term memory in Roscoe over some 30,000 DMTS trials distributed over 6 years. The numbers in parentheses are estimates of the total number of DMTS trials received up to that point. The data point on the right is based on 33 trials with a 9-min retention interval given one trial per day. (From D'Amato, 1973, by permission of Academic Press, New York.)

such as spatial cues, response bias, and the like (Sidman, 1969). We will not address this important task here. Instead, we will describe an avenue of research that has sought to link together animal STM and discrimination learning with delayed reward.

Inherent in the "temporal discrimination" hypothesis is the implication that, like other forms of discrimination training, performance in DMTS ought to increase with practice, which of course is exactly what we found. Conversely, there ought to be a limit to the improvement in DMTS performance that practice can confer, for, as with other modalities of discrimination, temporal discriminations cannot be performed with infinite precision. Therefore, under any set of fixed experimental conditions a limit ought to be reached with regard to the retention interval that can be endured by a given species of animal. For our monkeys this limit seemed to be in the order of a few minutes; despite years of practice, they were unable to match at a high level of accuracy when the retention interval exceeded 3 min. The 3-min limit could, however, be an artifact of our particular experimental conditions. In particular, the animals were subjected to a fairly large number of trials each day with virtually no intertrial interval, thus producing a sequence of difficult temporal discriminations. Suppose that only a single trial were given daily. The 3-min retention interval would then occur against the background of a 24-hr intertrial interval. Even granting that the comparison stimulus on the present trial had served as sample on the previous trial, would not the difference between the times to last seen as sample be so great as to be discriminable to the animal? Worsham (1973) actually ran such a study, with disappointing results. Only Roscoe, whose data for a 9-min retention interval appear in Fig. 3.1, was able to improve significantly his DMTS performance beyond the 3-min limit, and even so his performance at longer retention intervals was not particularly impressive.

It seemed clear to us that factors other than temporal discriminative ability were limiting our animals' DMTS performance and hence our estimates of their STM capacity. A prime suspect lay in the long retention intervals themselves. Waiting minutes in the experimental chamber for the opportunity to respond and perhaps receive reinforcement is apparently aversive to animals, not only to monkeys but to dolphins as well (Herman & Gordon, 1974). Evidence that long retention intervals are aversive takes a number of forms, such as extremely long latencies in setting up trials, undue amounts of "adjunctive" behaviors, and inattentive responding to choice stimuli. In order to counteract the untoward side effects of very long delays, we earlier adopted the procedure of randomly intermixing a sizable proportion of short-delay trials. Nevertheless, the occurrence of short-delay trials during a daily experimental session is no guarantee that disruptive emotional responses will not be elicited by long retention intervals and interfere with matching behavior on such trials. The question naturally arose as to what the "STM" of our monkeys would look like if we could find some way of neutralizing the aversiveness of very long retention intervals.

SHORT-TERM MEMORY AND DISCRIMINATION LEARNING WITH DELAYED REWARD

We could have approached the problem frontally by manipulating a variety of variables during the retention interval in the hope of finding a set of conditions that would simultaneously reduce the aversiveness of long delays and increase DMTS performance. Because of the possibility of introducing retroactive effects through retention interval manipulations, this approach seemed hazardous. Moreover, it might be difficult to discriminate whether a variable that facilitated DMTS performance did so because it reduced the aversiveness of long delays or because it somehow improved the retention process itself.

It appeared that a solution lay in the formal similarity that exists between two-choice DMTS and simple discrimination learning with delay of reward (DOR). In both situations certain information must be retained over a delay interval devoid of supporting cues. Perhaps the major difference is that in DMTS the information to be retained concerns the identity of a stimulus event, whereas in the DOR task it is the identity of the stimulus to which the subject has previously responded. Perhaps an animal subject is at an advantage in the DMTS paradigm, inasmuch as the response required at the end of the retention interval tends to call attention to the essential nature of the task. In DOR, in contrast, the connection between receipt of reinforcement and any information that survives the delay interval is not quite as apparent. However, with well-practiced animals in which the DORs are increased gradually, perhaps this difference too becomes inconsequential.

In any event we decided to shift to the DOR paradigm, employing a two-pronged strategy. First, if our analysis of the basic similarity between DMTS and discrimination learning with DOR was correct, our animals should be able to acquire simple discriminations with DORs of the same magnitude as they managed in DMTS (that is, 2 or 3 min). This would be no small accomplishment in view of the fact that discrimination learning in animals has been frequently shown to be greatly impeded by even relatively short DORs. Indeed, the deleterious effect of DOR on behavior acquisition generally has been a cornerstone assumption in the field of learning. We now know that this generalization does not hold for conditioned taste aversions (see chapter 2), but exceptions to the rule in discrimination learning are few, and for the most part, very recent.

Once we had demonstrated efficient discrimination learning with DORs of 2 or 3 min, the thrust of the research would shift to finding a set of experimental conditions which, operating during the DOR period, were capable of maintaining discriminative performance at extremely long delays, in the order of an hour or more. Presumably, such a set of conditions would be relatively benign with regard to inducing negative emotional responses during long periods of delay. Accordingly, the limits of our animals' STM capacity would then be reassessed, employing these very same conditions during the retention interval of DMTS. As things turned out, our research became impaled on the first objective for a period of more than $2\frac{1}{2}$ years.

Maintenance of a Simple Discrimination with Delayed Reward

Possibly animals have difficulty acquiring discriminations with DOR because (a) they are not adequately instructed as to what is required of them, and (b) without appropriate countermeasures, a substantial DOR may prove aversive, much like long retention intervals in DMTS. We therefore decided to train animals on a simple discrimination with immediate reinforcement; a short DOR would then be introduced, to be increased gradually until a DOR of 2 or 3 min was reached. Presumably at this point considerations (a) and (b) would both have been satisfied, and the animal would then be introduced to a new discrimination with a DOR of perhaps 2 min.

Five capuchin monkeys (*Cebus apella*), three of which had extensive prior experience on DMTS, were trained on a simple red-square discrimination with immediate reinforcement. The experimental apparatus has been described in detail previously (e.g., D'Amato, 1973). In brief, the discriminative stimuli were presented by inline projectors, faced with transparent plastic keys that served as the response mechanisms. There were five such projectors, one at each corner of a 12-cm square and the fifth at the center. A subject initiated a trial by responding on a microswitch 15 times, which resulted in S+ (red) and S− (square) appearing on two of the outer four projectors. A correct response resulted in delivery of one 190-mg Noyes banana pellet, while an incorrect response was followed by a 1-min time out. When participating in an experiment an animal is fed (an hour or later after an experimental session) rations sufficient to maintain its body weight at a level at which it works satisfactorily. The red-square discrimination was acquired rapidly, and we then introduced a .5-sec DOR, increasing the delay interval gradually. As was the case for the retention intervals in our DMTS paradigm, all DORs were spent in near-total darkness. A daily session usually consisted of 24 trials.

Although the deleterious effect of DOR on discrimination acquisition is well known, it is not generally recognized that discriminative performance can also be seriously impaired when a DOR (of only several seconds) is introduced abruptly into an already established discrimination (e.g., Mishkin & Weiskrantz, 1958). This was the basis of our precaution of increasing the DOR slowly. In spite of this measure, however, a number of the subjects suffered a striking deterioration in performance at only moderate DORs.

Figure 3.2 shows the results of three animals that prior to this experiment had experienced many hundreds of DMTS trials. Each point on the graph is based on a session of 24 trials, and sessions having the same duration of DOR are indicated by connecting lines. The best of the three, Olive, performed well under DOR until a delay of 30 sec was reached, at which point her performance steadily declined. It rebounded when the DOR was reduced to 4 sec and remained at a high level as the DOR was increased to 30 sec. Note, however, how Olive's performance steadily deteriorated at the 30-sec DOR, descending for three sessions to a level of only 66% correct.

Pete showed a persistent decline in performance the first time the 30-sec DOR was encountered, responding for three sessions significantly more to S− than to S+. After partial recovery at shorter DORs, Pete was continued on the 30-sec DOR in the hope that he would utlimately adapt to this delay interval. As the right half of his figure shows, after reaching a high level, his performance declined and remained erratic over the last 12 of these sessions.

Basil was particularly unable to cope with DOR. Deterioration of his performance set in at only an 8-sec DOR. He was able to maintain a high level of performance on the next cycle of increasing delays until 20 sec was reached. Note how once again performance successively declined with a constant DOR, as if the animal were learning not to respond to S+. After some maneuvering, Basil's performance reached a high level with a 16-sec DOR. However, it declined once more when the DOR was increased to 30 sec.

These data are puzzling. On the basis of the animals' performance at short DORs we must assume that the contingencies associated with responding to S+

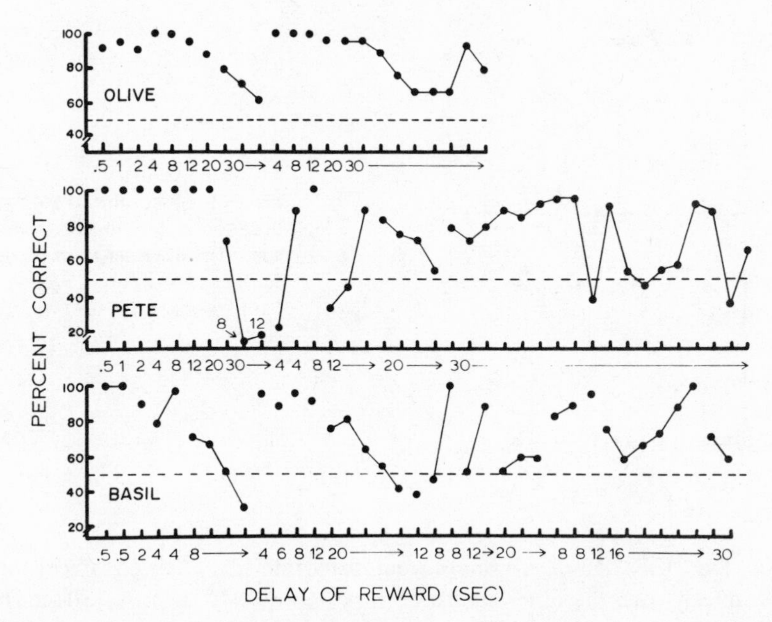

FIG. 3.2 Accuracy of performance on the red-square discrimination for three subjects with extensive past DMTS experience. Each data point is based on one session of 24 trials. Sessions having the same duration of DOR are indicated by connecting lines.

and S– had been well established. Why then did the animals increase their responding to S– at the longer DORs, when by this action they not only failed to receive a reward pellet but they suffered a 1-min time out as well? It is as though when the DOR becomes too much for the animal, it commences to "unlearn" the discrimination. This cannot be the entire story, however, because for one thing there are instances where with the DOR held constant the animal increased its responses to S– until it chose that stimulus more often than S+ (Pete and Basil). Second, when the DOR was reduced, frequently performance bounced back almost immediately, which would not be expected on the unlearning assumption. It appears that an emotional component is involved, perhaps in addition to a process such as unlearning. Apparently, long DORs are aversive to the animals, and because S+ is more frequently associated with the delay interval (because the animals respond more to S+ than to S–), this stimulus tends to be avoided. The emotional component can also account for the erratic behavior often observed with constant lengthy DORs. A shift in responding to S– only serves quickly to increase the conditioned aversiveness of this stimulus, which in turn promotes responding to S+.

Some support for this view is provided by the quite different results presented in Fig. 3.3. Edgar and Phurp are two laboratory reared animals with little or no previous experimental experience. Note how quickly Edgar reached the 30-sec DOR and note the minimal cyclicity in his performance over the nine sessions

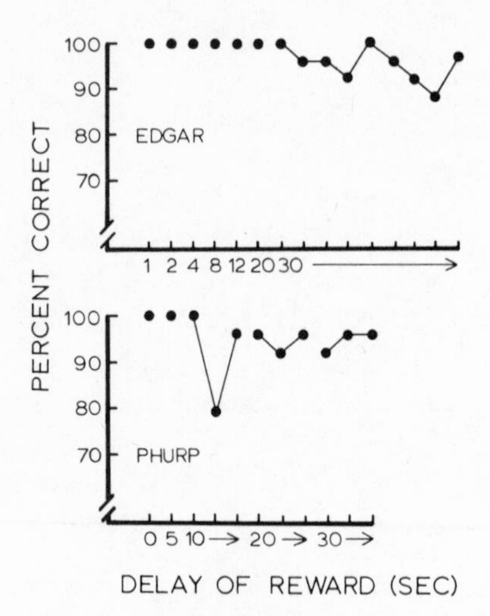

FIG. 3.3 Accuracy of performance for two subjects with very little prior experimental experience. The discrimination problem for Edgar was red vs square; for Phurp it was a dot (S+) vs a triangle (S–).

run at this DOR. Phurp was moved out from immediate reinforcement much more quickly than the other animals, but he too quickly reached a 30-sec DOR and only showed a small emotional effect in the process.

Although there is no reason to suppose that the DOR could have been increased indefinitely with Edgar and Phurp (see below), it nevertheless is clear that these animals had far less difficulty accommodating to the increasing DORs than did the first three subjects, which one might have supposed were in a favored position because of their extensive past experience with DMTS.

The implications we draw from the present data are as follows:

1. The aversiveness of the DORs was, in the case of Olive, Pete, and Basil, largely a result of the prior experience that these animals had with the response–reinforcer relationship, which was always immediate. In the absence of this prior extensive learning, capuchin monkeys are apparently disposed to tolerate moderate DORs fairly easily.

2. Extensive exposure to substantial retention intervals in the DMTS paradigm seems not to transfer to the response–reinforcer relationship. If the experienced animals had learned to tolerate the retention intervals of DMTS, this learning seems not to have had an important effect on the delay intervals of DOR discriminative performance. The reverse might not be true, however. Animals trained to endure substantial DORs might, because of that experience, be less disturbed by lengthy retention intervals in DMTS.

Preference for DOR- versus DMTS-Type Delays

In view of the susceptibility of the DMTS sophisticated animals to the disruptive effects of DOR on discriminative performance, the question arises of whether the delay period in DOR is more aversive than a comparable retention interval in

DMTS. In different terms, is a period of delay more aversive when it comes between the discriminative response and presentation of the reinforcing event, as in DOR, or when it precedes the response–reinforcement sequence, as in DMTS? To investigate this issue we allowed subjects to choose between two types of trial sequences, indicating their choice by pressing one of two levers located on either side of the projector display. In the "DOR-type" trial, as soon as the appropriate lever was pressed, S+ (red) and S– (square) appeared on two of the outer four projectors. When the animal responded to one of the discriminative stimuli, the delay period was entered. A reward pellet or a 1-min time out was delivered at the end of the delay for correct and incorrect responses, respectively. Pressing the other lever led to the "DMTS-type" trial, which differed from the former in that the delay occurred immediately on pressing the lever. At the end of the delay interval, S+ and S– were presented and the ensuing discriminative response immediately produced the food pellet or the time out. The DMTS-type trial, then, positions the delay period before the response–reinforcer sequence, as is the case for DMTS. In both types of trials the delay period was spent in near-total darkness.

Four animals from the previous study served in this experiment, and the general experimental procedure was similar to that of the earlier study. To acquaint the animals with the different contingencies associated with the two levers, on six of the daily 24 trials only the lever leading to the DOR-type trial was operative (indicated by an illuminated display above the lever), and on six of the trials only the lever leading to the DMTS-type trial was operative. Both levers were available on the remaining 12 trials. This procedure was instituted with a 1-sec delay interval and continued until the delay interval reached 20 sec, at which time both levers became available on all 24 trials. The lever associated with DOR-type trials was alternated between the right and left lever, as indicated in Fig. 3.4.

Focusing first on the animals' preference behavior, Pete and Basil showed an overall preference for DOR-type trials, at least until the second block of 40-sec delay trials was reached. Olive got caught up in a position preference (for the left lever) from which she did not escape until late in the experiment. However, Edgar showed an early preference for DOR-type trials but this faded at the 10-sec delay trials.

With two animals showing a preference for DOR-type trials and one enmeshed in a position preference, it is only fair to conclude that a delay interposed between the discriminative response and the subsequent (primary) reinforcement is at least as much preferred as a delay that precedes the response–reinforcement sequence. The present experiment is complicated, however, in that the initial lever-press response is itself involved in a response–reinforcer sequence, namely the conditioned reinforcement provided by the discriminative stimuli. It may be argued, therefore, that when the animals choose a DOR-type trial they are actually choosing immediate conditioned reinforcement over delayed conditioned reinforcement. However, it is almost impossible to avoid a complication

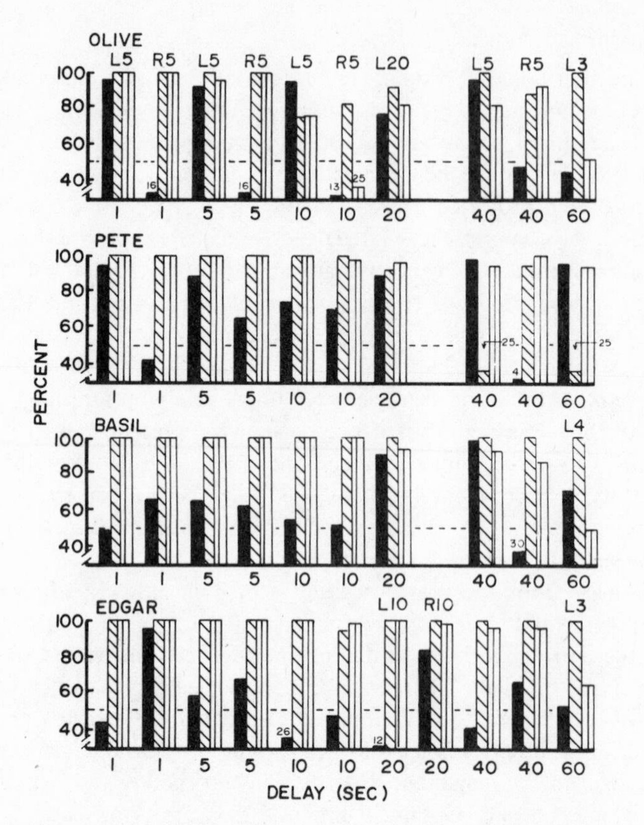

FIG. 3.4 The solid bars indicate the percentages of choice of DOR-type trials. The diagonally striped bars show the percentages of correct discriminative responses on DMTS-type trials, whereas the vertically striped bars show these percentages for DOR-type trials. Quantities less than 40% are written in. The position of the lever (left or right) leading to DOR-type trials is indicated on the top of the figure, as is the number of sessions under each condition. Except where indicated otherwise, the same conditions were in effect for all subjects.

of this nature, and in any case if a delay inserted between the discriminative response and primary reinforcement is particularly aversive it ought to override the contribution of the earlier lever response–conditioned reinforcer link.

Note in Fig. 3.4 that until the 60-sec delay was reached there was relatively little difference in the percentage of correct responses produced on DOR- and DMTS-type trials, performance in both cases remaining at a relatively high level. At the 60-sec delay, however, three of the four subjects (Basil, Olive, and Edgar) committed not a single error on DMTS-type trials, whereas their accuracy on DOR-type trials fell to between 50 and 65%. This difference in performance on the two types of trials is highly significant for all three subjects.

This last result shows unmistakably that at the time the discriminative stimuli appear, the animals are able to discriminate between the two types of trials and that their responses are under control of the prevailing delay condition. It is clear testimony that the animals' errors on DOR-type trials cannot possibly be caused by an unlearning factor. Errors on these trials were committed not because the animals had forgotten the contingencies associated with the discriminative stimuli, but because of the aversiveness of the long DOR period that ensued. Why then did they not choose DMTS-type trials and therefore avoid the delay of reward? Judging from the high performance levels at the 40-sec delays with both types of trials, it is likely that delays less than 60 sec were not sufficiently aversive to shift the animals toward the DMTS-type trials, which required foregoing the immediate conditioned reinforcement provided by the discriminative stimuli. Had the experiment been continued, it seems likely that a preference for DMTS-type trials would have developed.

The additional experience with DOR provided by the present study enabled all of the animals to maintain a high performance level on the red-square discrimination at 40-sec DOR, which represents a marked improvement over their performance in the first experiment. However, in view of the poor performance turned in by three of the four animals at the 60-sec DOR and the suggestion that the delay interval of a DOR trial might be more aversive than a comparable DMTS retention interval, it seemed unlikely that our monkeys could be made to cope with a 2- or 3-min DOR without the institution of special delay-interval conditions. Before we proceeded in this direction, we decided to lower our sights and get some evaluation of their ability to acquire new discriminations with DORs in the neighborhood of 45 sec.

Acquisition of Visual Discriminations with Delay of Reward

Three of the four animals that served in the previous experiment, Olive, Basil, and Edgar, were employed in the present study as well. As Table 3.1 shows (extreme right column), these animals differed widely with regard to their previous DMTS experience. Three additional animals that had no prior DOR experience were included in the experiment in order to get some idea of the importance of this variable. As with the "experimental" subjects, the "controls" varied considerably with regard to prior DMTS experience (Table 3.1).

Before we introduced the new discriminations, the experimental subjects were given 5–7 days of training on the familiar red-square discrimination with a 45-sec DOR. To help maintain the animals' motivation, in this and subsequent phases of the experiment the DOR was only 5 sec on from 25 to 50% of the red-square discrimination trials. All trials on the new discrimination were at the specified DORs. The control animals were run on the same red-square discrimination for 20 sessions (480 trials), all with immediate reward.

We thought it possible that acquisition of new discriminations with DOR would be facilitated in the experimental subjects if they were simultaneously exposed to delayed reward on the familiar red-square discrimination. Consequently, 12 of each daily 24 trials were devoted to the new discrimination and

TABLE 3.1
Results of Delayed Reward Acquisition Experiment

	Trials to criterion on new discriminations Delay of reward (sec)				Number of Previous DMTS trials
	1	15	30	45	
Experimental					
Olive	120	168	120	84	1,500
Basil	60	60	180	240+	5,000
Edgar	72	240+	240+	NR[a]	0
Control					
Dagwood	36	228	240	108	3,000
Peanuts	84	120	192	NR	500
Hubert	60	240+	NR	NR	0

[a]NR, not run at this delay.

12 to the red-square discrimination with the "long" DOR set at 45 sec. In the case of the control subjects, all trials on the red-square discrimination were always with immediate reward.

For each experimental subject, the stimuli that served as S+ and S− in the new discriminations were chosen quasi-randomly from a six-stimulus set consisting of vertical line, inverted triangle, dot, circle, plus, and a horizontal line. The same three sequences of discriminations used with the experimental animals were also assigned to the three control subjects. Acquisition of the first new discrimination took place with a 1-sec DOR. Subsequent new discriminations were given with DORs of 15, 30, and 45 sec. During the last of the new discriminations, the DOR on six of the red-square discrimination trials was increased to 60 sec and the DOR for the other six was 5 sec.

As in previous experiments, the subject's response to S+ and S− terminated the discriminative stimuli and initiated the DOR interval, which was spent in darkness. Therefore, there were no conditioned reinforcement cues whatever available. At the termination of the delay interval, a banana pellet was delivered on correct trials; a 1-min time out occurred instead on incorrect trials. The acquisition criterion was two successive days with 11 or 12 correct responses on the 12 new discrimination trials. In addition, an animal that reached criterion in less than eight sessions was required to maintain criterial performance through Session 8; failing that, its criterion score was advanced to the next occasion that its accuracy level was at least 92% for two successive days. The purpose of the latter requirement was to lessen the chance that our criterion would be satisfied by stimulus preference rather than bona fide learning.

Figure 3.5 presents the results from Olive and Basil. The first point of interest is that, although the performance of both animals on the background (old) discrimination was at a high level before the 1-sec DOR new discrimination was entered, acquisition of the latter was accompanied by a deterioration of the

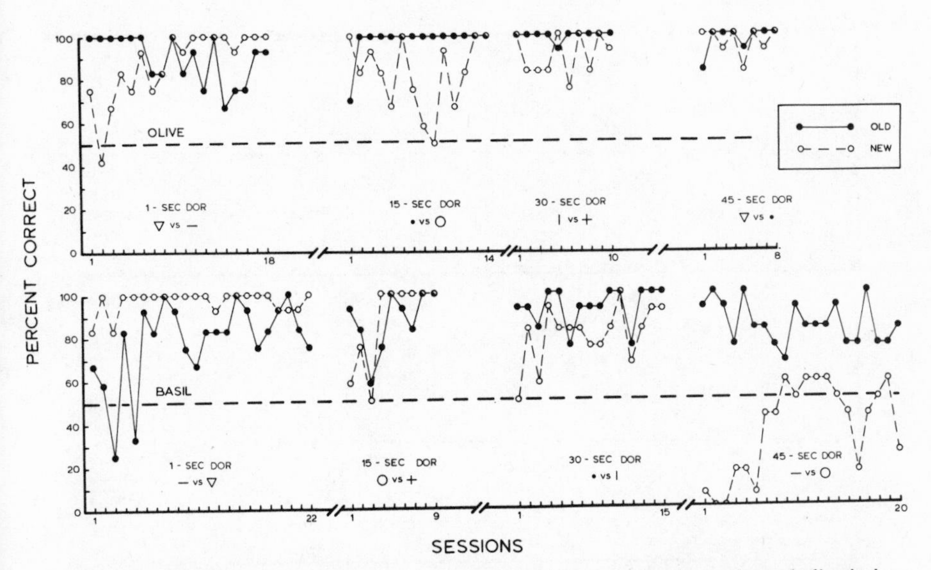

FIG. 3.5 Acquisition data for two of the three experimental animals on the new discriminations (open circles). The duration of the delay of reward and the identity of S+ and S– (S+ vs S–) are shown in the figure. Each session was comprised of 24 trials, 12 devoted to the new discrimination and 12 to the red-square background (old) discrimination (filled circles).

former, a kind of DOR contrast effect. Both animals went on to acquire the 15-sec and 30-sec DOR discriminations within the numbers of trials shown in Table 3.1. Because there were only six stimuli to choose among as discriminanda, the 45-sec DOR discrimination employed the stimuli that served as S+ in the 1-sec and 15-sec new discriminations. In an effort to bias the 45-sec DOR discrimination against the animal's stimulus preference, the initial trial on this discrimination (not counted in subsequent calculations) was incorrect irrespective of the subject's choice, and the stimulus chosen thereafter served as S–. Despite this handicap, Olive acquired the 45-sec DOR discrimination instantly, but because her accuracy dropped to 83% on the fifth session, 84 trials were required to meet our criterion. In the case of Basil, our method of assigning S+ and S– resulted in a strong bias against S+, which he was unable to overcome within the allotted 240 trials.

Edgar, the third experimental subject, was unable to reach criterion on the 15-sec DOR discrimination within 240 trials. At this point the DOR was reduced to 5 sec, and Edgar met criterion in 108 trials. Increasing the DOR to 15 sec did not disturb his performance, but he subsequently showed little learning on the 30-sec DOR discrimination over the allotted 240 trials.

With respect to the control subjects, Figure 3.6 and Table 3.1 show that Dagwood and Peanuts managed to meet criterion on all of the new discriminations to which they were exposed. (Because of ill health, Peanuts was not run on

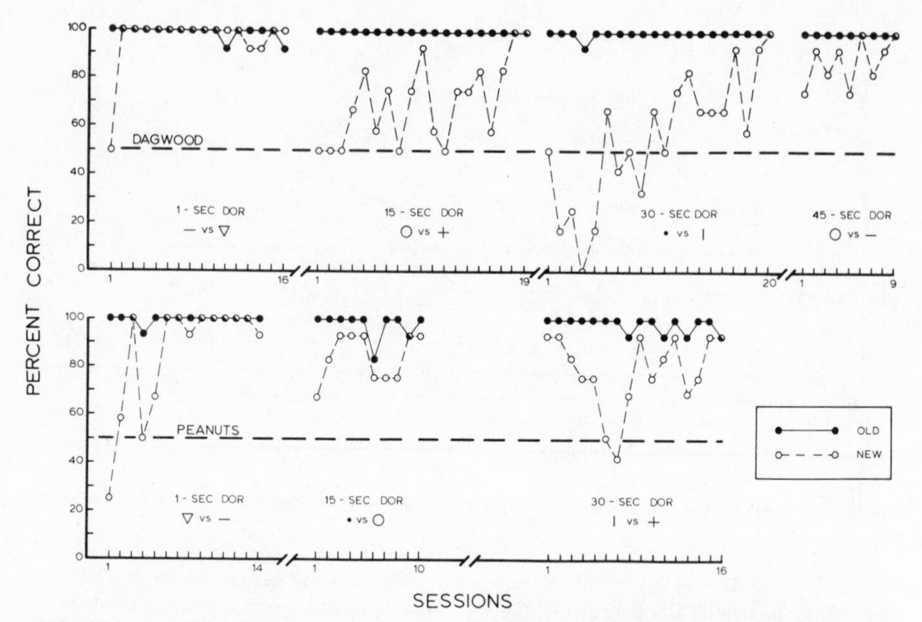

FIG. 3.6 Acquisition data for two of the three control animals on the new discriminations (open circles). The duration of the DOR and the identity of S+ and S– are given in the figure. Twelve of the 24 daily trials were on the background red-square discrimination with immediate reward (filled circles).

the 45-sec DOR discrimination.) Note that in contrast with the experimental subjects, the performance of the control animals on the background discrimination did not suffer as a result of acquisition of the 1-sec DOR discrimination.

The third control animal, Hubert, failed to acquire the 15-sec DOR discrimination within 240 trials. Still unable to learn the discrimination after an additional 240 trials with the DOR reduced to 5 sec, he was terminated. It is interesting that this animal, whose performance was the worst of the six subjects, had the benefit neither of previous DMTS nor of previous DOR experience.

Edgar, the one experimental subject that had no prior DMTS experience, did poorly on the new discriminations. Did this mean that prior DMTS experience was more important than DOR performance in facilitating discrimination learning with DOR? Apparently not. Phurp, whose earlier experience with DOR performance appears in Figure 3.2, was given an additional 24 sessions on the dot versus triangle discrimination, at the end of which he was able to maintain a high performance level at 90 sec DOR. He was then introduced to a 45-sec DOR new discrimination, with no background discrimination trials included. His performance reached a level of 90% correct responses in only seven sessions of 24 trials each (Fig. 3.7). Shifted to a 60-sec DOR new discrimination, Phurp's performance reached the same level by the seventh session, but then the cyclicity so characteristic of DOR performance with long delays made its

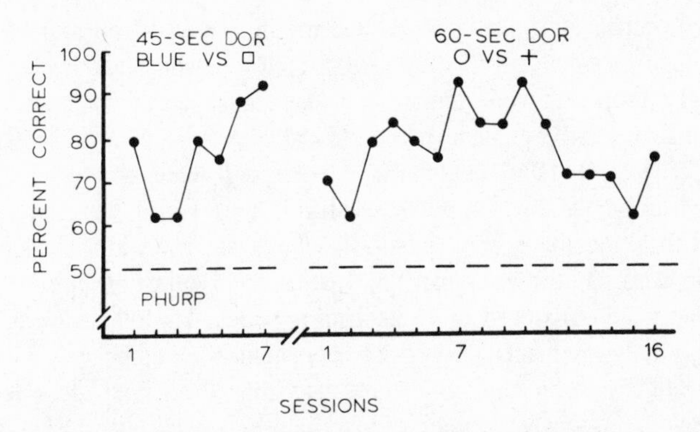

FIG. 3.7 Acquisition data for Phurp with DORs of 45 and 60 sec. Each session consisted of 24 trials on the new discrimination.

appearance. We interpret the decline in performance which occurred during the latter sessions to mean that the long DOR had become aversive to the animal.

Although the present results are not as definitive as one may like, certain conclusions seem warranted. First, our animals are clearly capable of reasonably efficient discrimination learning with DORs up to at least 45 sec. Prior experience with either DMTS or DOR performance may facilitate discrimination learning with delay of reward. The one animal, Hubert, that had the benefit of neither was totally incapable of learning a new discrimination with a DOR of only 5 sec, although this animal had had a fair amount of prior discrimination learning experience. Moreover, it is exceedingly unlikely that Phurp would have so readily acquired the 45-sec DOR discrimination were it not for his prior substantial experience with DOR performance. Taken together, then, the results offer the hope that if ways of coaxing the animals to endure far longer delays in discriminative performance could be devised, one might then show acquisition of new discriminations with much more impressive DORs.

Effect of Alcohol on Delay of Reward Performance

Alcohol has been shown to reduce the frustrative effects of partial reinforcement and extinction. If, as we assume, long DORs induce a negative emotional state similar to frustration, alcohol can be expected to facilitate DOR performance. We therefore ran a careful experiment in which the amount of alcohol ingested prior to a day's session of DOR trials was manipulated.

Edgar and Basil once again served as subjects, and at the time the experiment commenced they had amassed approximately 3,700 and 4,100 DOR trials, respectively, on the red-square discrimination. The DOR experience of the other two animals in the study, Hubert and Dagwood, was limited to that gained in the previous experiment. Once more the red-square discrimination was used, with the DOR adjusted to 60 sec over 14 or 15 pretraining sessions. The 60-sec DOR was chosen because at this value the performance level of all animals provided ample room for a facilitative effect of alcohol to manifest itself. The other experimental conditions were similar to those employed in the first experiment.

Unlike humans, monkeys have to be coaxed to consume substantial quantities of alcohol. The first 10 days of the experiment were therefore devoted to exploring a number of highly preferred fluids that could serve as vehicles for diluting and masking the ethanol alcohol, which was given in a concentration of 4.5–6%. The dosages studied were 0.5, 1.0, and 1.5 gm of ethanol per kilogram of body weight, administered in an ascending series. The alcohol was given 1 hr before the animal was placed into the experimental chamber, where it received 24 60-sec DOR trials on the red-square discrimination. Each dose level was in effect for five consecutive sessions. Five baseline sessions, during which the animals were given the fluid vehicle minus the alcohol, preceded and followed each dose level. All subjects ingested the total amount of assigned alcohol in all conditions, so there is no question about the alcohol getting into the animals.

Figure 3.8 summarizes the results of the experiment averaged over the four subjects. In a word, alcohol had very little effect on DOR performance, a conclusion that was verified by appropriate statistical analyses. Nor did any animal examined individually show an effect due to alcohol administration. However, the time required to complete the FR 15 on the microswitch, by means of which a trial was initiated, was greatly increased by alcohol, indicating that the drug was present in sufficient quantity to cause significant behavioral effects.

Note from Fig. 3.8 that the average level of performance during the first baseline condition was approximately 75%. Thirty-five sessions later, by which time the animals had accumulated an additional 840 DOR trials on the red-square discrimination, their performance had declined to a mere 63%. Clearly, DOR exerts a potent and lasting disruptive effect on discriminative performance.

Terminal Delay of Reward Performance

Figure 3.8 by no means represents the limits of DOR performance in our animals, for as with other behaviors, DOR performance tends to increase with accumulated practice. Nevertheless, only grudgingly did the performance limits imposed by DOR retreat as experience mounted. Figure 3.9 presents acquisition and performance data for Pete and Basil on a different discrimination problem (dot versus triangle). Basil's performance at the 60-sec DOR represents the highest level ever achieved by this animal at a DOR of this duration. When the DOR reached 90 sec, however, his behavior deteriorated badly. Pete, in contrast, maintained a high performance level through the 90-sec delay interval, and in later testing he managed 78% correct responses with a 120-sec DOR. It should be kept in mind, however, that these rather modest accomplishments represent for

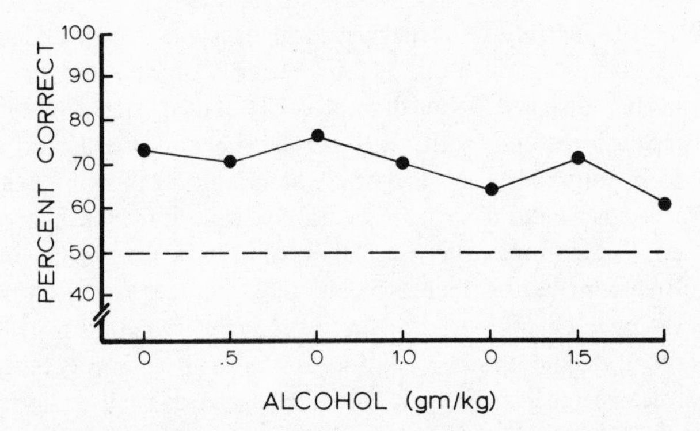

FIG. 3.8 Performance of the four experimental animals on the red-square discrimination with a 60-sec DOR. Each data point is based on five 24-trial sessions. The amount of alcohol ingested (per kilogram of bodyweight) prior to an experimental session is shown on the abscissa. Control sessions are indicated by 0 gm/kg.

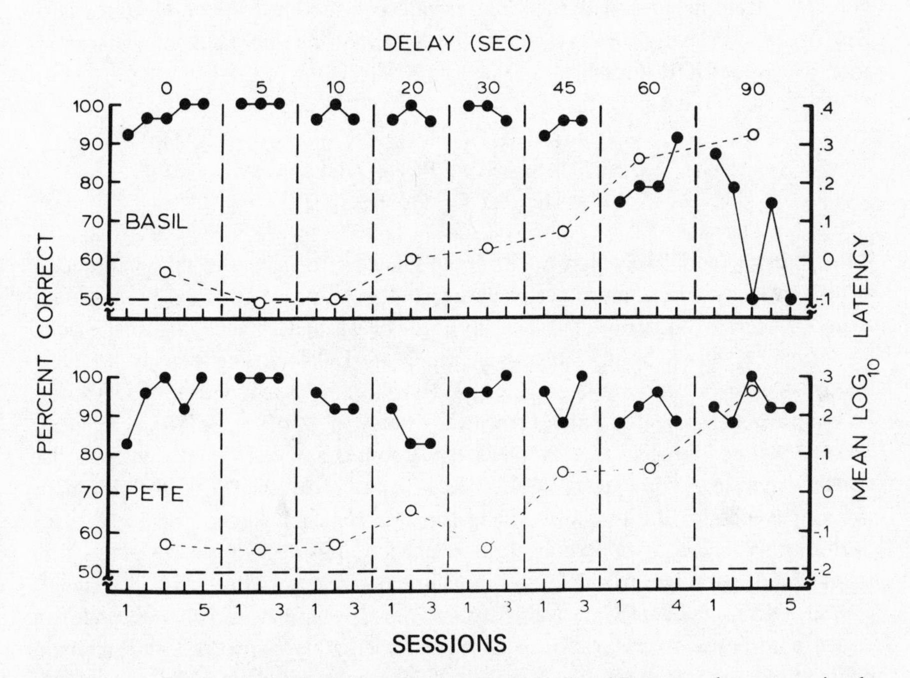

FIG. 3.9 The acquisition and performance results of Basil and Pete on a dot versus triangle discrimination with increasing DORs (solid lines). Each session is based on 24 trials. The dashed lines present the mean log latency of choice responses averaged over all sessions with the same DOR.

Basil and Pete, respectively, the culmination of some 7,300 and 4,900 DOR trials given over more than 300 and 240 sessions distributed over two long years.

We have so far discussed the disruptive effects of DOR only on choice behavior, because in a well-practiced discrimination task disruption of accuracy is perhaps more surprising and less understandable than increases in latency measures. We of course did observe a general increase in response latencies as the DOR increased. For example, the time taken to initiate a new trial–analogous to the postreinforcement pause–increased, as did the latency of a choice response (the time consumed between setting up a trial and responding to one of the stimuli). To illustrate the latter, Fig. 3.9 presents for Basil and Pete the mean log latency of choice responses for each DOR, averaged over all sessions conducted with the same delay. In both cases choice response latencies increased sharply once the DOR exceeded 30 sec. Interestingly, however, the increase in choice latency was correlated with a decrease in response accuracy in Basil but not Pete.

Observation of the animals over closed-circuit TV showed that with long DORs an animal often would set up a trial, approach the stimuli, start to respond to S+, then stop and withdraw from the stimulus display. Frequently several such abortive attempts preceded the final response, which not uncommonly was to S–. This sort of behavior reveals in a dramatic way the reluctance of the animals to enter a long DOR interval.

DISCRIMINATIVE PERFORMANCE WITH DELAY OF PUNISHMENT

As it did not seem likely that the limits of DOR performance could be extended significantly by additional practice alone, at this point we attacked the problem from a different direction. The best hypothesis we had to account for the poor performance shown by our animals at moderate DORs was that delay intervals are aversive, a common enough interpretation among investigators of DOR and its counterpart in related human research, "delay of gratification" (e.g., Mischel, 1974). By assumption, this aversiveness becomes associated with S+ and the animals therefore come to choose, at least temporarily, S–. Why not eliminate the aversiveness of the delay by giving the animals a reinforced task to perform during the delay interval? We attempted to accomplish this objective by shifting from a DOR paradigm to a delay of punishment (DP) procedure.

Basil, Olive, and Pete were trained on a new discrimination (plus vs circle) in which a response to either S+ or S– was immediately reinforced with a single pellet; however, responses to S– resulted in a substantial time out (3–5 min). This discrimination (based on avoiding the punishment of a time out) was acquired fairly rapidly. A delay was then introduced during which the display unit above the left lever was illuminated, indicating that a variable interval (VI) food schedule was available on this lever. Of course, the animals had already been shaped to respond to the illuminated left lever on a VI schedule. At this point, delivery of the food pellet at the end of the delay period was eliminated and incorrect responses continued to be punished by a 5-min time out. To recapitulate, when an animal responded correctly to the discriminative stimuli, it entered a period of delay during which a VI food schedule was available on the left lever; at the end of the delay interval the animal entered a short intertrial interval, after which it was free to initiate the next trial. After an incorrect response, the subject encountered the same VI food schedule during the ensuing delay interval; however, a 5-min time out was interposed between completion of the delay interval and onset of the intertrial interval. Because the same VI schedule followed correct and incorrect responses, no differential reinforcement for the discriminative response was available until the end of the delay period. The value of the VI food schedule varied from animal to animal and from time to time but was such as to allow from zero to four pellets on any single trial. Except for the illumination provided by the display unit above the left lever, the delay interval was spent in darkness.

We assumed that because a fairly generous VI food schedule was available during the delay interval and the delayed consequence was negative rather than positive, the delay interval could in no way be aversive to the animals. It is true that immediate shock is often preferred to delayed shock, both in rats and humans (e.g., D'Amato & Gumenik, 1960; Knapp, Kause, & Perkins, 1959), suggesting that a delay interval interposed between an instrumental response and a noxious event can be aversive, most likely because of the anxiety generated by the delay. In the present situation, however, the delay interval was filled with a rewarded activity instead of being empty, and a time out was used as the aversive event instead of a noxious electric shock. It is hard to believe that under these conditions anticipation of the terminal aversive event can generate enough anxiety during the delay interval to neutralize the positive affect from the VI food schedule. As we shall see later, this appraisal is supported by the animals' choice latencies.

A total of from 76 to 79 DP training sessions, distributed over a 3-month period, was received by each animal. The data that are presented here are based on sessions which, because of the long time outs employed, were limited to a maximum of 90 min, rather than being fixed at the customary 24 trials. The top panel of Fig. 3.10 presents the values of the DP, VI, and time out parameters over the final 26 sessions of Basil's training. After considerable maneuvering, Basil was finally able to maintain a reasonably high level of performance with a 60-sec DP. When the delay was increased to 90 sec, however, his performance fell off precipitously. The last 16 sessions were devoted to an attempt to improve his performance by increasing the duration of the time out to a final value of 20 min and by decreasing the VI to 20 sec, but these efforts failed.

Olive's performance on the terminal 30 sessions of DP training are shown in the middle panel of Figure 3.10. This animal, which earlier was able to maintain

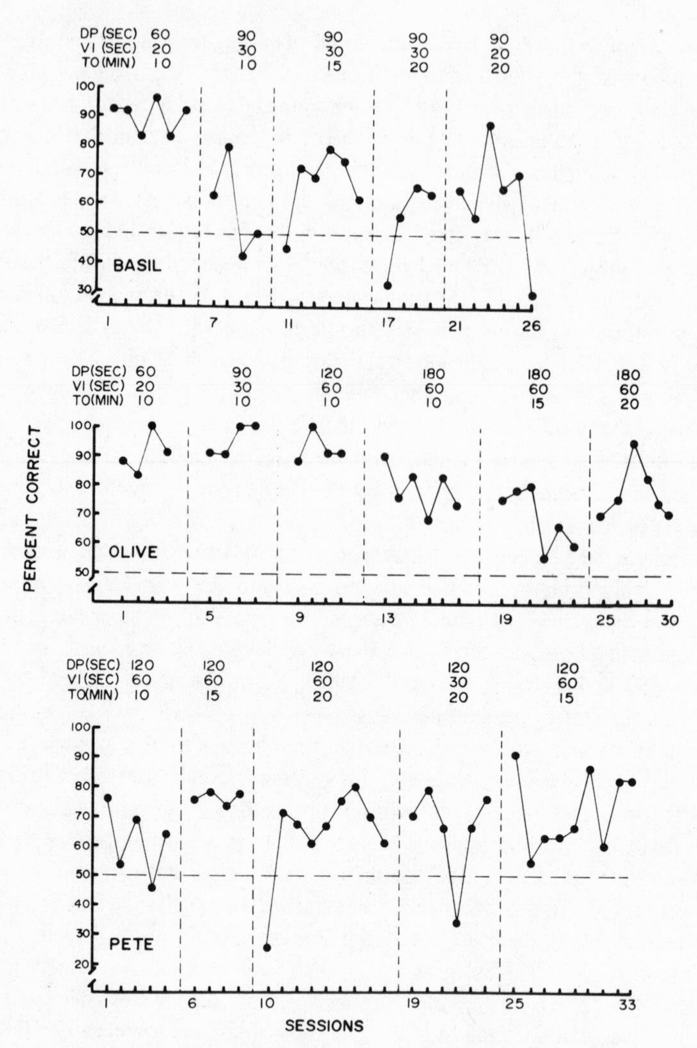

FIG. 3.10 Terminal discriminative performance of Basil, Olive, and Pete on delay of punishment (DP). The values of the DP, variable interval food schedule (VI), and time out (TO) parameters are given at the top of each graph. Each session was run until the animal had completed 24 trials or 90 min had elapsed.

a 60-sec DOR only with great difficulty, performed at a very high level even at a 120-sec DP interval. However, her performance declined when the DP reached 180 sec, and increasing the timeout from 10 to 20 min did not help.

The bottom panel of Fig. 3.10 shows Pete's performance over 33 sessions of 120-sec DP, during which the VI and time out parameters were manipulated in an effort to maximize his performance. It is evident from the figure that, although certain combinations of conditions produced better performance than others, no set of conditions was able to maintain performance at a high level of accuracy.

We calculated for Basil and Pete the mean log latency of choice responses for DPs ranging from 10 to 60 sec. Unlike the latencies obtained with increasing DOR (Figure 3.9), there was no increase whatever in choice response latency with increasing duration of the DP. Moreover, the DP latencies were comparable to the shortest latencies obtained with DOR. These findings support the view that, in contrast to DOR, the delay intervals of our DP task are not aversive.

DISCUSSION

Needless to say, the results we obtained during our more than 2 years of research on DOR and DP were not at all what we had expected. We began our work believing that with sufficient training of the proper sort our monkeys ultimately would become capable of successful DOR performance with delays in the order of an hour or more, and having accomplished this, they could perhaps be shown capable of substantial DMTS performance with retention intervals of the same order of magnitude. Obviously our accomplishments fell far short of this goal, despite a sustained attack on the problem from a variety of directions. We have by no means described all of the avenues that we explored; let us merely observe that over the course of the research Basil, Pete, and Olive, the most intensively studied animals, had the benefit of some 12,000, 8,000, and 7,000 total trials, respectively, on DOR and DP, distributed over approximately 540, 350, and 300 sessions. The results of this intensive and prolonged training were that Basil ultimately became capable of responding at a reasonably high level of accuracy with a 60-sec delay. For Pete and Olive we may say that the limits of delay were pushed to 120 sec or so. Moreover, inasmuch as DP training followed the DOR phase, it is not clear that the animals' performance limits are different for the two paradigms.

Effects of Delay of Reward on Performance Speed and Accuracy

In relating our results to the relevant literature, we will focus first on the effects of DOR on performance, distinguishing between measures of speed and accuracy. As mentioned above, introducing a DOR, even gradually, usually results in disruption of response speed. To cite only two illustrations, Bullock and Richards (1973) reported an enormous increase in the postreinforcement pauses of a pigeon responding on an FR 50 as the DOR was increased from 0 to 30 sec. The time to complete the FR 50, the work time, also increased, as did the variability of both of these measures. Ferster and Hammer (1965) trained rhesus monkeys and baboons to endure a 24- and 18-hr DOR, respectively, and found that at these extremely long delays the animals often required 10 hr or more to complete an FR 50 even though an enormous reward was used, equivalent to one or more day's full ration. From these and many other studies one may conclude that time-based measures of response performance, such as response rate or speed, are greatly affected by DOR which, when long enough, leads to a dramatic slowdown in the performance of even well-practiced behavior and a substantial increase in variability. Moreover, these disruptive effects often cannot be circumvented simply by increasing the magnitude of reward or by extensive practice.

Turning to the effects of DOR on performance accuracy, we find that the available data are very sparse. Mishkin and Weiskrantz (1958) reported a sharp reduction in the accuracy of visual (successive) discrimination performance in rhesus monkeys when a DOR of only 8 sec was introduced abruptly. In the only other directly relevant study known to us, Lawrence and Hommel (1961) reported quite different results. They trained rats on a black–white simultaneous discrimination with a 10-sec DOR and, after a 90% criterion was reached, some of the animals were tested with the DOR increased successively to 20, 30, and 60 sec. Although the authors reported accuracy data only for the 20-sec DOR, at which the animals maintained criterion performance, they implied that accuracy did not deteriorate at the longer DORs. However, they explicitly noted that at the longer DORs the animals hesitated in the start box, often for 30 sec or longer before making a choice. The failure to find a drop in accuracy with increases of the DOR to 60 sec may be because the animals had no prior experimental experience with immediate reinforcement; indeed, even the original discrimination was learned with a 10-sec DOR. That only 50 trials were given at each DOR during the discrimination performance phase may also be relevant. In any case, however, our data provide extensive evidence to show that performance accuracy on a visual discrimination can suffer greatly when the DOR is only in the neighborhood of a minute or so. Although the point at which DOR disrupts performance accuracy can be advanced by increasing practice, there seems to be a distinct limit to this process.

Possibly the simplest interpretation of the deleterious effects of DOR on performance speed and accuracy is to view the delay period as an aversive stimulus. Responses that produce the delay therefore tend to be suppressed, just as in other punishment situations, allowing other responses (grooming, exploring, etc.) to emerge, and this is reflected in longer response latencies, lower rates of responding, and longer postreinforcement pauses. When an alternative, experimentally defined response is available, as in a simultaneous discrimination task, the effect of the response-produced aversive stimulus is to shift the animal from one response alternative to the other (that is, from S+ to S–).

The increase in variability that occurs both in performance speed and in accuracy measures can be accommodated in the following way. Whether the alternative response engaged in is experimentally defined or not, it tends to be suppressed not only by the DOR but also by the additional delay it occasions. In our situation, for example, the alternative behavior of responding to S− results in the DOR plus a 1-min time out and postpones the opportunity for reward until the next trial. In the case of a simple instrumental response, the behaviors that occur during a lengthy postreinforcement pause are presumably suppressed in some measure by the fact that they postpone the next reward for a duration longer than the programmed DOR. When alternative behaviors are sufficiently suppressed, the animal promptly returns to the instrumental behavior itself. The resulting short response latencies combine with the previous long latencies to generate the observed high variability of performance. A similar process presumably occurs with regard to performance accuracy, but here runs of correct and incorrect responses, instead of short and long response latencies, alternate.

As to the reasons a substantial DOR should take on aversive characteristics, our results are suggestive. Past experience with immediate response–reinforcer relationships in the same or similar experimental situations seems to exacerbate the disruptive effects of DOR, perhaps as a kind of contrast effect. Even without such experimental training, however, the ordinary experience of an animal in his everyday environment generates a whole range of response–reinforcer relationships in which reinforcement is immediate or nearly so. The animal advances to the water bottle and almost immediately drinks the water it seeks. It reaches for, grasps, and soon ingests the food pellets in its vicinity. Consequently, even without explicit prior experimental training, immediate response–reinforcer relationships are the norm, and where exceptions exist, such as in the restoration of water equilibrium after fluid intake, there are ample conditioned reinforcers to fill the gap. Presumably, this extensive background experience with immediate response–reinforcer relationships establishes a generalized expectation that, when violated by a long DOR, results in negative emotional responses.

Delay of Reward and Reward Utility

In addition to its role as an aversive stimulus, DOR may also influence behavior through its effect on the incentive value or utility of the reward object (Renner, 1968). The incentive value of a reward is by no means determined solely by its physical characteristics, and in addition to such variables as the subject's motivational state, no doubt delay is an important factor. Receipt of $10 here and now is generally preferred to receipt of the same amount 6 months hence even though there may be no question regarding the certainty of the later delivery. When we turn to more objective evidence, we find it has been reported for rats in a two-choice situation that a logarithmic increase in amount of reward is required to compensate for a unit increase in delay (Davenport, 1962). Consequently, apart from any effect from emotional factors, a DOR tends to reduce the value of a reward object in much the same way as a reduction in reward magnitude. This property of DOR needs to be kept in mind when its impact on behavior is considered, for even if the aversiveness of the delay were somehow completely eliminated, some disruption of performance is likely to be observed.

Perhaps the effect of the aversive component of a DOR is simply to reduce further the utility of the reward object. According to this point of view, the negative utility of the aversive component combines at each point in time with the positive utility of the reward object to form a "net" utility that governs behavior (Renner, 1967).

Foresightful Behavior and Delay of Consequences

The assumption that DORs act as aversive stimuli and reduce the utility of the reward object goes a long way toward explaining the deleterious effect of delay of reward on measures of performance speed and accuracy, but how are we to explain the rather comparable effect exerted by delay of punishment (DP)? Aversiveness of the delay period can hardly be a factor. Moreover, it seems unlikely that a DP of 1 or 2 min can reduce the aversiveness of a 20-min time out until it is only moderately less preferable than no time out at all. Before we pursue this issue, we wish to place it in a broader context.

A possible interpretation that may occur to many is that the disruption of performance caused by DOR and DP is a result of a memory deficit. The animals fail to make the connection between their discriminative behavior and the subsequent differential reinforcement because, at the time of reinforcement, information regarding the nature of their last response is no longer available. It is interesting, from the point of view of this hypothesis, that DOR and DP performance are not very different from DMTS performance in terms of the delay intervals that can be tolerated. On the other side of the ledger, the results reported above for DOR- vs DMTS-type trials do not support such an interpretation.

In any case, however, we prefer to look at the delay deficit in a more general way. One of the reasons that we have reservations about the heavy reliance placed on human models of memory as prototypes for animal memory is the disregard this approach carries for ethological and evolutionary considerations. We seem already to have forgotten how our view of learning was broadened by incorporating ethological and evolutionary factors into our analysis. Very likely, the same will prove true for memory. Several of the general findings of animal memory become more comprehensible when we concern ourselves with the question of what memory is good for (from a point of view of survival). For example, it is valuable, if not essential, that animals be equipped to remember for long periods of time the location of the vital necessities of life. Consequently, it should not surprise us that chimpanzees are capable of remembering the location where food was deposited for as long as 48 hr under conditions of testing that resemble a short-term memory task (Yerkes & Yerkes, 1928), and it becomes reasonable that animals generally perform far better on spatial delayed-response tasks than on the nonspatial variety. The observation that bees are apparently capable of remembering the location of a food source for more than 30 days (Lindauer, cited by Eibl-Eibesfeldt, 1970) also makes sense. Because an animal which forgets the meaning of cues that signal a stalking predator is not likely to get the opportunity to relearn their significance, we may expect aversive events to generally prove unusually resistant to forgetting. Although there are some suggestive data on this score (cf. Hoffman, 1969), this interesting research question has received far less attention than it deserves.

Now that we have come to accept the concept of species-specific learning, given the intimacy of learning and memory, can we do less for species-specific memory? Where survival requires or is enhanced by the preservation of specific information for long periods of time, we should suspect the existence of specialized mechanisms to perform this function. To mention only two possible illustrations of such specialized memories, preferences established by imprinting are often usually resistant to forgetting, and certain strains of salmon and trout are apparently able to remember characteristics of their home stream for 4 years or more. How are these instances, and others like them, of highly specialized and remarkably functional memories to be understood in terms of current models of human memory, based as they are on an information-processing system that lies, as it were, at the end of the evolutionary line?

An ethological–evolutionary approach may also help make sense of the present results. Performance under delay of consequences, whether DP or DOR, may be thought of as a form of "foresightful" behavior. The animal is required to choose between two behaviors the differential consequences of which are not visible until some time later. What role in survival, we may ask, does foresightful behavior of this sort play in the life of monkeys? Not much, probably. Capuchin monkeys do not plant seeds for crops that do not become visible for weeks or perhaps months. They do stimulate branching of trees and therefore the amount of future foliage by their feeding habits, but there is no indication that this is performed with any foreknowledge of the terminal consequence (Oppenheimer & Lang, 1969). It is interesting that where foresightful behavior is required for the well-being or the preservation of the species, specialized "preprogrammed" mechanisms have evolved, as in nest building in birds and hoarding in rodents. The deposition of heavy fat layers prior to hibernation may also be thought of as a preprogrammed foresightful behavior.

A significant observation from our laboratory–which, obvious though it was, went unnoticed for years–indicates how indisposed toward foresightful behavior *Cebus* monkeys can be. Each of our animals is fed once a day with a ration of monkey chow biscuits that is in excess of the animal's daily need (if it is not on deprivation), which has generated the daily chore of sweeping up the biscuits discarded by the animals. Hours before feeding time, the animals show signs of hunger and when the food becomes available, they grab at the biscuits and eat them voraciously. Shortly, however, they squander their excess supply by allowing the biscuits to fall through the mesh floor into the pan below or, just as frequently, by throwing them out of the cage to the floor of the room where they are irretrievably lost. Although most of our animals have been in laboratory residence for at least 9 years, during which time they have experienced deprivation for extended periods, not a single animal has ever been observed to develop the foresightful behavior of saving present excess biscuits in its food bin, or elsewheres, for later use. Possibly, then, the failure of our animals to learn spontaneously to conserve food for later use is related to their poor performance with delayed consequences, in that both may result from the lack of selective pressure for foresightful behavior.

How does the present hypothesis fit in with the large number of studies showing that animals are capable of maintaining behavior, albeit at reduced efficiency, with substantial DORs, as long as 24 hr? Before we answer this question, we need a clearer idea of what is meant by foresightful behavior. On reflection we seem to use that term to refer to behavior maintained by consequences the utility of which is currently very low but which increases as a function of time. As an illustration, when our monkeys are fed their daily food ration, the utility of food biscuits presumably drops to a very low value as a function of satiation, subsequently increasing as satiation wanes. If they developed the behavior of saving excess biscuits when satiated, this would constitute foresightful behavior.

The utility of a reward object can also be reduced to low levels by delaying its occurrence sufficiently long. Not everyone is inclined to contribute weekly to a Christmas Club for the distant reward that instrumental behavior makes possible. In such situations the utility of the reward presumably increases as the time to its realization approaches. Neglecting for a moment the potential aversiveness of the delay interval, prompt performance of instrumental behavior that results in a reward object which, because of DOR, currently possesses a very low incentive value is also a case of foresightful behavior. The contribution of the DOR's aversiveness may be simply to depress the utility of the reward below the level that can be managed by the delay interval alone, as we have pointed out earlier.

With this interpretation of foresightful behavior, its existence in any particular instance is not necessarily an all-or-none affair. Animals are probably particularly disinclined to engage in foresightful behavior when the utility of the reward object is reduced to near zero because of satiation, although there is virtually no experimental evidence on this point one way or the other. In the case of loss of utility caused by DOR, animals may sometimes cease responding altogether for delayed food even when it constitutes their only source of sustenance (Ferster & Hammer, 1965), but this is not the usual result. More often instrumental behavior is disrupted, the extent determined by the duration of the delay. Can we say in these instances that the animals are displaying some degree of foresightful behavior? Perhaps. Suppose we knew by some means that the utility of a 10-pellet reward delayed for 30 sec was generally equivalent to that of one pellet delivered immediately. If the performance level of an animal working with a 10-pellet reward under a 30-sec DOR were only at the level generated by a one-pellet reward given immediately, its behavior would not be considered foresightful because it was governed by the current utility of the contingent reward. If it were at the level appropriate for a two-pellet immediate reward, it would qualify as a case of foresightful behavior because of the control exerted by the future utility of the reward object.

The implication of this analysis is that maintenance of instrumental behavior at reduced efficiency in a DOR situation does not by itself constitute foresightful behavior. To take the Ferster and Hammer (1965) study as an illustration, after about a year of training, their two rhesus monkeys were able to sustain performance on a moderate FR with a 24-hr DOR. However, the time required to complete the FR–apparently comprised of mainly the postreinforcement pause–was characteristically in excess of 16 hr and not infrequently it extended to a day or more. We interpret these results to mean that soon after delivery and consumption of a reward–which was huge–owing to satiation the utility of reward dropped to a very low value. If the animals were capable of a high degree of foresightful behavior, they would nevertheless have performed the instrumental response in a relatively short time and therefore have limited the delay of the next meal to 24 hr. Instead, many hours had first to elapse in order to allow their motivation for food, and consequently the current utility of the future reward, to increase to a level sufficient to generate the instrumental behavior. Our conclusion, therefore, is that maintenance of instrumental behaviors with long DORs is not necessarily evidence of foresightful behavior. It may simply reflect control by the current utility of the reward object.

A major difficulty with this analysis arises when one tries to distinguish experimentally between control by current and by future reward utility. One may question whether there is any need at all to refer to future utility, which, one may argue, must in one way or another express itself in current utility. Although there is merit in this view, one can point to numerous examples where behavior seems to be motivated not by the associated current utility but by the future consequences of the act. We swallow an unpleasant medication because of possible future beneficial results, but this anticipation does not render the medication any less distasteful. It is precisely because not everyone is capable under such circumstances of acting in terms of future utility that flavor-disguised medications exist. A person who purchases an end of season sale article, such as an outdoor grille, an overcoat, or whatever, is acting in part on the item's anticipated utility in the coming season rather than on current utility, which is likely to be low. Possibly such cases can ultimately be recast in terms of current utility only, but at the present time the future referent seems an essential ingredient of foresightful behavior.

We have attributed our animals' relatively low performance levels at long DORs to their poorly developed capacity for foresightful behavior, that is, to their tendency to respond in terms of present rather than future utility. The utility of the reward object at the time of choice was, in turn, assumed to be reduced to a low level by a long DOR, in part because of the aversiveness of the delay interval. Possibly a similar analysis can be applied to our delay of punishment task, even though the delay interval cannot be assumed to be aversive. That the utility of an aversive event is negative rather than positive raises the possibility of symmetry between the delay interval effects of DOR and DP. If aversiveness of the delay in the DOR paradigm reduces the positive utility of the contingent reward, it is perhaps reasonable to assume that the delay interval of our DP paradigm—possessing positive affect because of the VI food schedule—acts to reduce the negative utility of the contingent time out. To take a somewhat remote illustration, one may be willing to risk heartburn as a result of dining in a fine French restaurant but not as a possible consequence of visiting a pizza parlor. Stating the point in a more testable fashion, a mild electric shock delivered with a 10-sec delay on one side of a T maze is likely to be avoided less by a hungry rat if both sides of the maze are baited with 10 food pellets than if they are empty. One implication of this interpretation for our DP studies is that the richer the VI that fills the delay interval, the less the control exerted by the time out. Unfortunately, because the DP, VI, and TO variables were frequently manipulated jointly and because the VI parameter was not varied widely, it was not possible to evaluate this implication properly. Nevertheless the hypothesis advanced seems a reasonable one, and it enables us to deal with the DOR and DP results in a consistent way.

Discrimination Acquisition with Delay of Reward

We have provided clear evidence that our animals are capable of reasonably efficient discrimination learning with DORs of at least 45 sec. Anything that limits the level of attainable discriminative performance likewise places limitations on discrimination acquisition, and it is therefore unlikely that our experimental conditions would support discrimination learning with DORs beyond a minute or so. In contrast, Lett (1974) has reported nonspatial discrimination learning with a 1-min DOR and cue-correlated spatial learning with delays as long as 1 hr (Lett, 1973, 1975). There are several features of Lett's procedures that probably have served to facilitate long delay learning. First, a small number of trials were given each day, in some experiments only one. Second, high-incentive rewards were employed, an entire day's food ration in the 1973 studies, and 2.5 ml of 25% sucrose solution in the later reports. Finally, and perhaps most important, the animals spent the delay period in their home cages rather than in the experimental apparatus. It is this last factor to which Lett assigns her success in demonstrating learning with long DORs. With Revusky (1971) she believes that by removing the animal from the apparatus "interfering associations" (competing responses) are prevented, thus allowing the animals to associate responding to S+ with the delayed reward.

Our interpretation of the effect of removing the animal from the apparatus during the delay period is that this operation serves to reduce the aversiveness of the delay period. Although similar to the Lett–Revusky hypothesis, ours is more specific in identifying the nature of the interfering responses. Our analysis also provides a rationale for the role played by the very large or highly preferred rewards employed by Lett. Because of the long DORs, the reward object must be extremely large in magnitude or very highly preferred, or both, for it to generate even a small incentive value (utility) at the time of choice.

In order to provide a basis for the facilitating role of the type of trial spacing used by Lett, we must return to the temporal discrimination hypothesis, discussed in the first section of this chapter. We have suggested that DMTS and discrimination learning with DOR involve similar processes both with regard to the aversiveness generated by the delay intervals and the demands on retention. In the case of DMTS we assumed that the retention component was based on a temporal discrimination process, and very likely a similar discrimination is involved in DOR learning. In order that reward and nonreward reinforce responding to S+ and S− selectively, the animal must be able to discriminate at the time of reinforcement whether it has last responded to S+ or to S−. This amounts to forming a temporal discrimination, similar to that assumed to be required in DMTS.

It is interesting in the light of this analysis that Lett's training procedures were likely to promote the establishment of such temporal discriminations. Recall that in several experiments the first correct response terminated the day's trials. Associating nonreward with responding to S− was thereby greatly facilitated because a previous rewarded trial (responding to S+) never occurred in closer juxtaposition with responding to S− than approximately 24 hr. That more than one response to S− must have frequently occurred in succession also aided associating nonreward with S−. Terminating a day's trials with a correct response should have facilitated associating reward with responding to S+, because only a single transition between responding to S− and to S+ was thereby allowed each day. Another factor to be considered is that the type of trial spacing employed by Lett produced an asymmetrical intertrial interval. The intertrial interval after incorrect responses was very short, only a matter of seconds. After correct responses, in contrast, the intertrial interval generally was approximately 24 hr. These greatly different intertrial intervals can serve as differential time markers that set off and make more distinct the events that precede them, in this case responding to S− or to S+ (D'Amato, 1973). The practice of spacing trials one per day, followed in some of Lett's experiments, should also greatly facilitate forming the temporal discriminations required by DOR learning.

Although the training procedures adopted by Lett were all to their advantage, her animals usually reached acquisition levels of only 75–85% correct responses. It should also be pointed out that not all attempts to demonstrate long-delay learning using Lett's procedures have been successful (Roberts, in press).

Nevertheless, given the proper experimental conditions some degree of discrimination acquisition with DORs in the order of an hour is very likely within the capabilities of animals. Most probably our monkeys fell far short of this range both in discrimination acquisition and performance because less than optimal conditions were employed. With regard to discriminative performance, requiring the animal to remain in the experimental chamber during the delay no doubt allowed a great deal more aversiveness to be developed by the delay interval than if the animal had been removed from the apparatus; the utility of the reward object at the time of choice could have been increased greatly if a much larger reward or a highly preferred reward had been used instead of the single 190-mg Noyes pellet. Finally, the large number of daily trials employed by us no doubt also was a factor, particularly in discrimination acquisition, where the formation of temporal discriminations was so important.

There is scant doubt that if these variables were optimized our animals' discriminative performance level would improve considerably, and as a matter of fact we recently obtained preliminary data that justified this assumption. However, can the limit of effective discrimination performance and acquisition with DOR be advanced indefinitely by procedural modifications, however well optimized? We think not. Although the limit may be pushed to an hour, or perhaps even to a few hours, infrahuman animals apparently do not possess the mechanisms that permit the maintenance of current behaviors, let alone the acquisition of new behaviors, when the contingent consequences of these behaviors are far removed in time. This ability seems to rest on the development of language, from which man has fashioned his time-preserving aids, backward and forward, that increase enormously his capacity for memory and foresightful behavior, respectively.

ACKNOWLEDGMENTS

The research described in this paper was supported by National Science Foundation grant GB 40111.

4

Studies of Short-Term Memory in the Pigeon Using the Delayed Matching to Sample Procedure

William A. Roberts
Douglas S. Grant

University of Western Ontario

HISTORICAL INTRODUCTION

In many respects, the delayed-response experiment with animals is analogous to the short-term memory (STM) experiments with human subjects devised by Brown (1958) and by Peterson and Peterson (1959). In both cases, the subject is presented with a stimulus that signals the correct response or response to be rewarded, the stimulus is removed from the perceptual field, and a measured period of time elapses. At the end of this period, the experimenter tests the subject's ability to make the appropriate response. Comparing the percentage of correct responses on a number of trials with some chance baseline indicates the goodness of retention at a given time interval, and a plot of percentage correct against different time intervals provides a curve of retention. Although the two situations are then similar, it is interesting to see that the courses of research on these two problems followed very different lines. In the 16 years or so since Brown's and the Peterson's experiments, STM in people has been analyzed intensively in terms of factors responsible for retention and forgetting. By contrast, the substantial research on delayed response done in the first half of this century provided little in the way of theoretical analysis of processes of memory in animals (Roberts, 1972a; Winograd, 1971).

To understand how this has happened, one must look at the conceptual beginnings of research on delayed response. The earliest laboratory experimentation on delayed response was done by two graduate students at the University of

Chicago under the direction of H. A. Carr. Although this work never reached completion, it provided the impetus for the classical studies by W. S. Hunter (1913). Hunter's primary reason for performing experiments on delayed response was to determine whether or not symbolic processes could be found in nonhuman organisms. He systematically rejected previously offered evidence that animals possessed ideas on the grounds that in each instance the behavior in case could be explained in terms of a response to a perceptually available stimulus. What was needed was a situation in which an animal's behavior could be attributed only to the mental representation of a previously presented stimulus. To Hunter, the delayed-response problem offered the situation *par excellence* for eliciting this behavior. Pieces of apparatus were devised in which an animal could view the illumination of one of three exits but was not released to choose among these exits until sometime after the illumination was terminated. Choice of the previously lighted exit led to reward, but choice of one of the unlighted exits failed to yield reward. Three species of animals, rats, dogs, and raccoons, were tested and, although all of the animals tested showed some ability to make the correct response after delay, the limits of delay varied widely among species; the maximal delay for rats was only 10 sec, whereas raccoons could delay for 25 sec and dogs for 5 min. Equally as important as these figures were the qualitative observations made of the animals' behavior during the delay interval. The success of rats and dogs appeared to be completely dependent on the maintenance of a positional orientation toward the correct exit during the delay interval, but raccoons were able to choose the correct exit regardless of their orientation during the delay. Hunter concluded from his experiments that a form of sensory thought was used by raccoons. It was theorized that an intraorganic cue, that is, an idea, was formed by the lighting of an exit but waned during the delay and was rearoused at the time of release, acting to elicit the correct response. In the case of rats and dogs, however, the capacity for such representative factors was held to be absent, these animals being completely dependent on the maintenance of bodily orientation as a mechanism for bridging the delay.

The work of Hunter was to have a long-lasting influence on subsequent research in delayed response. Hunter had suggested that species differed in their capacities for symbolic representation and that the delayed-response problem could be used to reveal these differences in two ways. The first way was through observing whether or not an animal required orientation in order to bridge the delay. A clearcut qualitative distinction could be made by observing this behavior–species that were capable of ideational thought did not require orientation, and those not capable of such thought did require orientation. The second way in which species could be differentiated was by measuring the limits of delay. Clearly, species without symbolic processes should not be able to delay successfully for as long a period as species with symbolic processes, particularly if the possibility for use of orientation was reduced to a minimum. Furthermore, among species possessing symbolic processes, differences in the degree to which these processes were developed might be revealed by maximal delay time. Therefore, species could be ranked in terms of their delay time, providing a phylogenetic hierarchy of the evolution of symbolic capacities. Indeed, Hunter's procedure—gradually incrementing the delay interval until performance reached chance—suggested a mental testing approach to delayed-response experimentation.

In the 30 years or so following the publication of Hunter's monograph, considerable research was addressed to the problems he had raised. Cats, rats, dogs, and a variety of primates were tested to determine whether or not they required orientation to perform delayed response and to determine the limits of their delay capacity. The outcome of these experiments could only be viewed as disappointing for Hunter's original thesis. A number of techniques were devised to prevent animals from orienting toward the correct stimulus during a delay, and yet all of the species studied, even the rat, showed successful delayed response under these conditions. In the face of considerable evidence that animals could delay response successfully in a number of situations where orientational cues were ruled out, many investigators came to the conclusion that all of the species so far studied could bridge the delay through use of a centrally stored memory trace. Attempts to scale animals in terms of delay limits also failed. The problem here lay in the fact that no absolute limit could be found for any given species. Variation in method, particularly the use of spatially separated and visually differentiated stimuli, led to a constant upward revision of the maximal time for which an animal could delay. It became clear that type of apparatus and experimental procedure were far more important factors in determining the limits of delay than was species.

Although some early investigators saw delayed response as a test of memory, it was seen primarily as a potential device for comparing the intelligence of species. Most of these early experiments simply were not designed to analyze processes of memory in animals. Only in the last few years have researchers begun to view delayed-response techniques as a means of studying the basis of memory in animals, undoubtedly in large measure a result of the recent explosion of interest in human memory. As a result, various delayed-response procedures have been used recently to study STM in pigeons (Grant & Roberts, 1973; Roberts, 1972b; Roberts & Grant, 1974; Shimp & Moffitt, 1974; Zentall, 1973), rats (Roberts, 1972c; Roberts, 1974), and monkeys (Jarrard & Moise, 1970; Jarvik, Goldfarb, & Carley, 1969; Moise, 1970; D'Amato, 1973).

The major purpose of this paper is to review and discuss a recent series of experiments carried out by the authors on STM in the pigeon. The procedure used in these experiments has been delayed matching to sample (DMTS), a variant of the delayed-response problem in which only the visual characteristics of a stimulus and not its spatial position must be retained in order to respond accurately after a delay. Our approach to this work contrasts with that of the early research discussed above. We have not been concerned with the limits of delay in the pigeon or with using this measure to compare the pigeon with other species in terms of intelligence. Instead, we have concerned ourselves with an attempt to discover factors that promote retention and forgetting in pigeon STM. Drawing partially on studies of human memory and partially on our own intuitions, we have chosen certain variables that we have felt may have significant effects on STM in animals and have studied their effects on DMTS. The objective of this approach has been to formulate a theory or set of theoretical principles governing STM in the pigeon. Such a theoretical structure then affords a base for the formulation of further experiments. In addition, this approach may have a comparative purpose. That is, parallel experiments may be done with other species, and the theoretical principles derived from these experiments may be compared across species to make inferences about similarities and differences in mechanisms of memory.

In the following sections, we shall consider the effects on pigeon STM of length of exposure to a sample stimulus, length of delay interval, and degree of spacing between repeated presentations of a sample stimulus. Furthermore, we shall report some experiments that investigated proactive interference (PI) effects in pigeon STM. A theory of STM in the pigeon is presented as it has evolved in the course of our studies.

THE EFFECTS OF SOME BASIC VARIABLES ON SHORT-TERM MEMORY

The Procedure

All of the experiments that are reported here have been carried out in operant chambers designed for testing pigeons. One wall of a chamber contained three pecking keys aligned in a horizontal row and spaced 8 cm apart, center to center; these keys were located at about the height of a pigeon's head. Projectors mounted behind these keys could present either colored fields or black and white patterns as stimuli. Below the center key was an aperture through which a bird could obtain reward from a lighted magazine containing grain. Reward for a correct response consisted of a 2- or 3-sec period of access to grain. The pigeons used in these experiments were kept at about 80% of their free-feeding body weight and so were highly motivated to work for reward.

Prior to our experiments, birds were shaped first to peck the three keys and were trained then on simultaneous matching to sample. Typically, the subjects were given 40–50 trials of simultaneous matching to sample per day until a criterion of either 85% or 90% correct responses was reached over a block of 5 days. On a simultaneous trial, a stimulus, say a red field, would appear on the center key. When the pigeon pecked this key, the side keys would become illuminated, one with the matching color, red, and the other with a nonmatching color, say green, and the center key would remain illuminated with red. A peck on the red side key produced immediate reward, but a peck on the green key would turn off all keys without reward. In some experiments, an incorrect choice terminated the trial, and in other experiments a correction procedure was used in which the keys were reilluminated after a 6-sec blackout following an incorrect response; opportunities for correction continued until the subject made the correct response and was rewarded. Intertrial intervals ranging from 10 to 120 sec in different experiments followed either reward or an incorrect response, following which the center key would be reilluminated for another trial. Of course, each stimulus was sometimes the sample and sometimes the incorrect choice from one trial to the next, and the left–right position of the correct and incorrect side keys was balanced across trials. In general, the mean number of days it took a group of pigeons to learn simultaneous matching to sample ranged from 15 to 25 days.

Once a subject had reached criterion on simultaneous training, the memory aspect of this research was introduced by turning off the sample stimulus on the center key and introducing some delay interval before the matching and nonmatching stimuli were presented on the side keys, with the center key remaining darkened during the side key presentation. Therefore, pigeons could not match to sample on the basis of a physically present stimulus but were required to do so on the basis of memory. The delays used usually spanned a period of 10 sec or shorter, and, in most experiments, a 0-sec delay was used in which the side keys were illuminated immediately after the offset of the center key. The operant chamber in which an animal was tested was kept darkened throughout testing, except for the illumination provided by the stimulus keys and the reward aperture.

Effects of Exposure Time and Delay

One variable that has been shown to have considerable effect on human STM has been length of exposure or presentation time of a stimulus to be remembered. Using the Peterson procedure, Hellyer (1962) found that increasing the number of repetitions of consonant trigrams from one to eight led to increasingly higher levels of retention and slower rates of forgetting. In free recall of word lists, it has been found that the length of time for which items are presented is related monotonically to number of words recalled, in a linear fashion in some experiments (Murdock, 1960), and according to a negatively accelerated curve in other experiments (Roberts, 1972d; Waugh, 1967). In human subjects, then, opportunity to process information is a fundamental factor in determining performance on a retention test. In light of this strong effect of presentation time on human STM, we were led to inquire initially as to whether the pigeon's memory might be affected similarly by length of exposure to a sample stimulus.

In an initial experiment (Roberts, 1972b, Experiment 1), the pigeon's exposure to a sample stimulus was varied by manipulating the fixed ratio (FR) requirement on the sample key; ratios of FR 1, FR 5, and FR 15 were used. The stimuli presented on sample and side keys were fields of blue, green, yellow, or red light. As an orthogonal variable, retention under each level of FR was measured at delays of 0, 1, 3, and 6 sec. It was expected that retention would decline across delays, as in other studies of DMTS. However, we were particularly interested in the way in which degree of exposure to the sample might interact with delay. Ten Silver King pigeons were tested repeatedly under all combinations of these two variables.

The outcome of this experiment is shown in Fig. 4.1, where percentage of correct responses, that is, choices of the sample stimulus, is plotted against delay for each FR with each point representing 720 observations or trials. Level of performance was enhanced substantially by increases in the FR requirement on the sample key; in fact, each animal tested showed a perfect ordinal correlation between FR and correct responses. Negatively accelerated forgetting curves are found at each level of exposure. Interestingly, these curves do not interact but fall in parallel to one another, showing about as much difference between FR values at the 6-sec delay as at the 0-sec delay. Hellyer's (1962) curves for human subjects are clearly different from these in that repetition affected rate of forgetting.

The curves presented in Fig. 4.1 suggest that the relationship between exposure duration and accuracy is negatively accelerated in that the increase in performance between FR 5 and FR 15 is about equal to that between FR 1 and FR 5, indicating that the contribution of each peck to correct responding decreases as ratio increases. One difficulty with this conclusion is that the length of time the pigeon is exposed to the sample stimulus is not strictly under the experimenter's control. If the pigeon fulfills the FR slowly, the presentation time is longer than if the FR is completed rapidly. It is possible, then, that a negatively accelerated relationship between FR and performance can result, even though presentation time and matching accuracy are related linearly. This could happen if the pigeon's rate of pecking increased as responding on an FR progressed. In order to eliminate this complication, we introduced a procedure in which pigeons were trained initially on simultaneous matching to sample with increasingly higher FRs on the sample key (up to FR 20) and then were transfered to a timer-controlled procedure. Under this procedure, the sample stimulus was presented for a preset fixed period of time on each trial. A trial always began with the onset of a white light on the center key, which the subject had to peck once in order to introduce the sample stimulus. Data from an experiment using this procedure are presented in Fig. 4.2 (Roberts & Grant, 1974, Experiment 1) for ten Silver King pigeons. The sample stimulus (a red or green field) was presented for .5, 1, 2, 4, or 8 sec and was followed by a delay of 0 or 1 sec. The curves shown in Figure 4.2 are based on 360 trials per point.

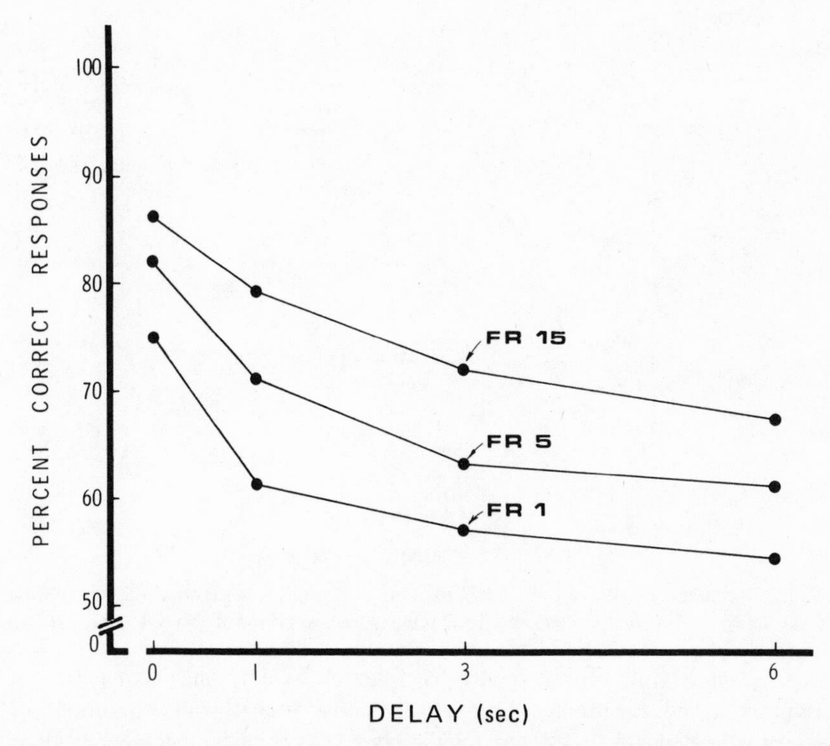

FIG. 4.1 Retention curves for three frequencies of presentation of the sample stimulus. (From Roberts, 1972b; Copyright 1972 by the American Psychological Association. Reprinted by permission.)

They show that accuracy increased generally as a negatively accelerated function of presentation time, although the small increases from .5 to 1 sec give them a slightly sigmoid appearance. Performance was consistently lower with a 1-sec delay than with a 0-sec delay, somewhat more so at the longer presentation times than at the shorter ones.

The presentation of matching and nonmatching stimuli on the side keys in a DMTS task represents a recognition test of memory. One way of accounting for recognition memory in this situation is to assume that a subject compares the stimuli presented on the side keys with a trace of the sample stimulus stored in memory. If a match between the trace and a test stimulus is made, a correct choice results. A critical factor determining the accuracy with which this match can be made is the strength of the memory trace; that is, the stronger the trace, the more frequently an accurate match will be made. Given this theoretical assumption, the curve relating DMTS accuracy to presentation time may be held to reflect the negatively accelerated growth of a memory trace with increased exposure to the sample stimulus. In addition, the decline in matching accuracy found with increasing delay may be attributed to a decay process, which is held

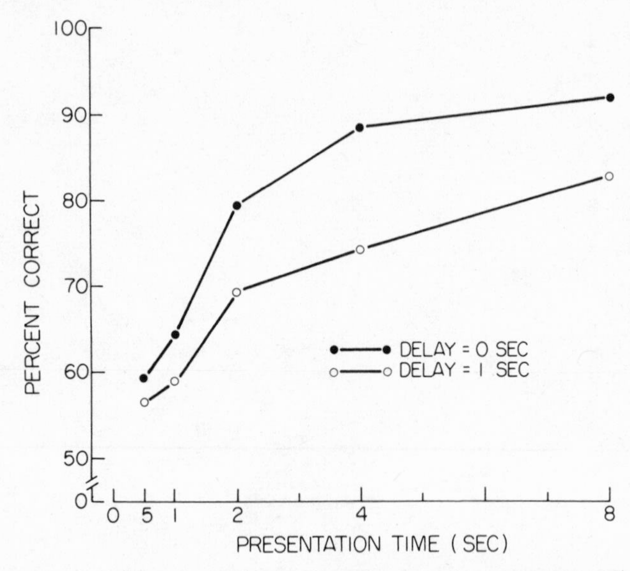

FIG. 4.2 Accuracy of delayed matching to sample plotted as a function of presentation time at delays of 0 (filled circles) and 1 sec (empty circles) (From Roberts & Grant, 1974.)

to take place as a negatively accelerated function of time since termination of the sample stimulus. Although this relatively simple strength and decay model of memory is inadequate to account for the complexities of human memory (e.g., see Tulving & Bower, 1974), it may be successful in explaining a number of the phenomena associated with the pigeon's short-term retention of sensory stimulation. Indeed, such differences in theory may be revealing of comparative differences in memory systems between species.

It should be noted that the theory proposed here makes no provision for memories of sample stimuli entering a long-term memory state. An alternative to the trace strength theory is an attributes model of memory (e.g., Bower, 1967; Spear, 1971; Underwood, 1969). This type of theory holds that the pigeon stores a memory of the sample stimulus in terms of a number of dimensional values or attributes that characterized the stimulus presentation. Presentation of the sample stimulus on a side key would then reactivate these attributes and lead to correct recognition. The effect of presentation time could be explained by the accumulation of an increasing number of attributes with prolonged exposure to the sample, and the delay function could be accounted for by the loss of short-lasting attributes necessary for a high level of immediate retention (Tulving, 1968). However, some attributes of the memory would be relatively permanent and could promote long-term retention of that stimulus event. Certainly, evidence for such a model can be found in studies of primate STM. D'Amato and his colleagues have reported that highly practiced monkeys can retain memory of sample stimuli in DMTS over periods as long as 9 min

(Worsham & D'Amato, 1973). Furthermore, retention in the monkey becomes independent of presentation time, a flash of the sample stimulus for a fraction of a second yielding as good retention as a more extended presentation (D'Amato & Worsham, 1972). Both D'Amato (1973) and Mason and Wilson (1974) have suggested that the monkey's forgetting in DMTS tasks may not represent a loss of memory for the visual characteristics of the sample stimulus but a failure to discriminate the point in time at which it has occurred. For example, a monkey presented with a cross pattern sample to remember and shown a cross and a circle as test stimuli has difficulty in remembering which of these two stimuli appeared most recently because both stimuli have appeared several times before as samples and distractors on preceding trials. This theory leads to the prediction that retention on any given trial should improve as the frequency with which the sample stimulus has appeared as a distractor on preceding trials decreases and as length of time since the last appearance of the sample as a distractor increases. Similarly, retention should improve as the frequency with which the distractor on the current trial has appeared as a sample on preceding trials decreases and as the time interval since its last occurrence as a sample is lengthened. Several experiments lend support to this theory. Mason and Wilson (1974) varied the frequency with which sets of stimuli appeared within a fixed period of testing on DMTS and found that the performance of rhesus monkeys decreased as frequency of presentation increased. In a similar experiment, Worsham and D'Amato (1973) tested monkeys on daily DMTS sessions of 24 trials and varied the sample set size used at any one session among values of two, three, and seven stimuli. Because the number of trials was fixed, the frequency with which stimuli were used as samples and distractors necessarily increased as the sample set size decreased. As predicted by the temporal discrimination theory, matching accuracy decreased as set size decreased. In another experiment, Worsham and D'Amato (1973) created easy and hard trial sequences for monkeys, "easy" and "hard" being defined in the context of temporal discrimination theory. On hard sequences, the distractor stimulus of a test trial had been the sample on the immediately preceding trial, whereas on easy sequences, the distractor of the test trial had not appeared as a sample for three or four preceding trials. In agreement with the notion that sample and distractor stimuli in hard sequences should be difficult to discriminate with respect to point of occurrence in time, hard sequences showed a lower level of performance on test trials than did easy sequences. One last piece of evidence comes from an experiment reported by Bessemer and Stollnitz (1971, Experiment IV). Learning-set sophisticated rhesus monkeys were trained on a concurrent discrimination learning procedure in which each of 16 problems was presented one time each day for a period of 4 days. A consistent improvement in correct choices was found over the 4 days of training. On the grounds of this observation and others, they concluded that "the mechanism of rapid intraproblem learning evidenced in learning-set performance appears to involve a dynamic memory process which is superimposed upon the gradual and constant increments produced by a passive memory process [p. 32]."

We feel that the evidence now available weighs against the idea that pigeons carry memories of DMTS trials for extended periods of time. Evidence to support this contention comes from a recent pair of experiments carried out by Grant (1975a). Four Silver King pigeons that had experienced approximately 16,000 trials of prior DMTS training were tested on alternating red–green and blue–yellow DMTS trials, with delay varied within days and sample presentation time varied between days. In Experiment 1, animals were given 60 trials per day, and presentation times of 1, 4, and 8 sec were used. Delay times were longer than those previously used with the pigeon, being 0, 10, 20, and 30 sec. In Experiment 2, there were 32 trials per day, and both presentation times and delays were lengthened; presentation times were 1, 4, 8, and 14 sec, and delays of 0, 20, 40, and 60 sec were used. Both experiments yielded essentially the same results, and retention curves for Experiment 2 are presented here in Fig. 4.3. Each point on these curves is based on 720 trials. The curves show that extended practice has raised the pigeon's level of matching at long delays; at a 60-sec delay, the level of matching is significantly above chance with presentation times of 4, 8, and 14 sec. However, the pattern of the influence of the variables studied remains the same as that found in earlier studies with less highly practiced birds. That is, we see a set of negatively accelerated forgetting functions that fall in parallel, with matching accuracy being a monotonic and negatively accelerated function of presentation time. Statistical tests showed that the effects of both delay and presentation time were significant and that these variables did not interact with one another. Although pigeons' DMTS performance is improved by practice, typical forgetting curves are observed with these animals, and performance does not become independent of sample presentation time, as is the case with highly trained monkeys (D'Amato & Worsham, 1972).

Of particular importance in Grant's study was an analysis of performance on each trial, trial n, as a function of that trial's relationship to the preceding trial using the same pair of stimuli, trial $n - 2$. In this analysis, performance was compared between the condition in which the roles of the stimuli on trials $n - 2$ remained the same and the condition in which the roles were reversed. For example, if red were the sample and green the distractor on trial n, then red would have been the sample and green the distractor on trial $n - 2$ in the same role condition, and red would have been the distractor and green the sample in the reversed roles condition. The analysis revealed no significant effect of these two relationships in either of the experiments; when performance was averaged over the conditions of presentation time and delay, the same and reversed roles conditions yielded matching scores of 66.8% and 66.2%, respectively, in Experiment 1 and scores of 72.5% and 69.8%, respectively, in Experiment 2. Two further analyses were performed. One involved the similarity of the colors used

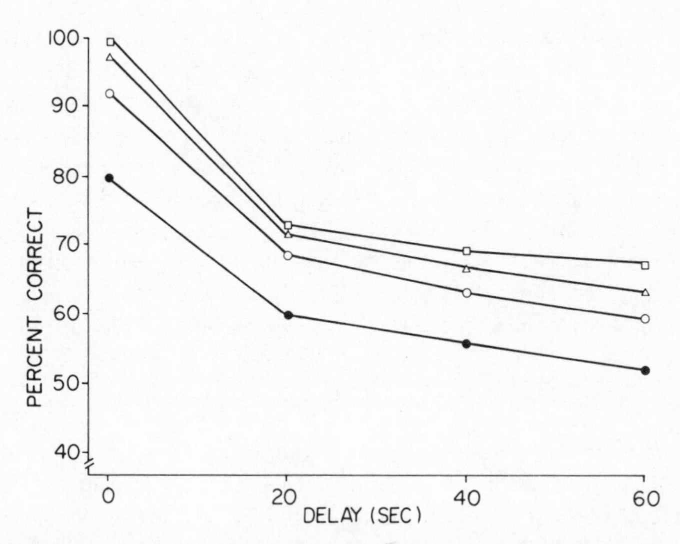

FIG. 4.3 Retention curves for delayed matching to sample when the presentation time of the sample stimulus was (•), 1; (○), 4; (△), 8; or (□), 14 sec. (From Grant, 1975a.)

on trial n to those used on trial $n - 1$. Here, again, same role and reversed role conditions were compared, with red and yellow being equated as one similar pair and blue and green as the other. No significant effect of the relationship between stimuli was found in either experiment. Finally, a third analysis compared performance on trial n for conditions in which the spatial position of the correct choice on trial n was the same or reversed from trial $n - 1$. Again, no significant effect was found in either experiment.

In another series of experiments, Grant (1975b) was able to demonstrate proactive inhibition (PI) in pigeon STM. This was accomplished by giving pigeons two successive DMTS trials, with the stimuli used as sample and distractor reversed in roles from the first trial to the second. It was found that this procedure increased the rate of forgetting on the second trial relative to a control condition in which no preceding trial was given. This effect appeared strongly if the second trial immediately followed the first but disappeared if a 20-sec interval was interposed between trials. Therefore, what evidence there is for sequential interference effects in pigeon STM suggests that such effects are temporally short lived. (A more extended treatment of PI effects is presented in a subsequent section of this chapter).

To summarize, the experiments done on the effects of exposure duration of the sample stimulus and length of delay with the pigeon all seem to be interpretable in terms of a theory proposing that the strength of a memory trace grows with increasing exposure to the sample stimulus and decays with the passage of time in the absence of the sample stimulus. Furthermore, the analysis of performance on DMTS trials preceded by other trials designed to provide conflicting information gives little evidence of long-term interference effects and suggests that long-term memories of events occurring on DMTS trials are not established in the pigeon. When the pigeon data are compared with those found in primates, it is tempting to suggest that quite different memory systems may be at work in bird and monkey. The monkey data suggest a memory system in which the encoding of stimulus information into memory can be accomplished completely in a brief instant and in which memories, once encoded, are relatively permanent. As usual, the validity of this suggestion awaits further research, but the contrasts already observed between species suggest that a comparative analysis of memory mechanisms may be a fruitful line of research.

Effects of Temporal Spacing of Sample Stimulus Presentations

One of the most intriguing phenomena found in studies of human memory has been the spacing effect. If an item to be remembered is presented to a subject twice before a test of memory, retention is higher if the two presentations are spaced apart in time than if they are massed. Furthermore, the facilitative effect of spacing increases with the length of the spacing interval, up to a certain limit. The spacing effect has been reported in experiments using presentation of a single stimulus as in the Peterson procedure and in experiments using presentation of multiple items, and the effect has been found with both recall and recognition tests of memory. For reviews of this area of research, see Bjork (1970) and Melton (1970).

The spacing effect has been viewed as puzzling or paradoxical because the time interval following the first presentation of an item leads to forgetting of that item and yet enhances retention if the item is repeated. Several accounts of the spacing effect have been offered, with the leading contenders being the notions that spacing either allows extra time for processing or rehearsal of information (Rundus, 1971) or broadens the contextual encoding of information (Madigan, 1969). In several experiments we have carried out with the pigeon, the effects of spaced repetition of the sample stimulus have been studied. We did these experiments for two reasons. Given the somewhat surprising effects of spacing on human retention, we inquired as to whether this effect was uniquely human or whether it might be found in other species. Second, studies of spacing provide an excellent test of the trace strength and decay theory we proposed to account for the effects of presentation time and delay. This theory clearly predicts an adverse effect of spacing, for any blank interval between sample presentations should lower the overall memory trace strength at the time of test. These reasons are related, of course, in that support for the trace strength and decay theory necessarily suggests a difference in the effects of spacing on human and pigeon STM.

The most illustrative study of the effects of spacing on STM in the pigeon is an experiment done by Roberts and Grant (1974, Experiment 2) in which four variables were studied in a factorial design. Essentially, a typical trial in this experiment required the pigeon to peck an initially white center key to produce either a red or green field as a sample stimulus; following this first presentation of the sample (P1), an interstimulus interval (ISI) in darkness followed, and the ISI was followed by a second presentation of the same sample stimulus (P2). After a delay, the side keys were lighted with red and green to test the animal's retention. The variables studied were length of P1, 1 and 4 sec; length of the ISI, 0, 2, and 5 sec; length of P2, 0, 1, 2 and 5 sec; and length of the delay, 0 and 2 sec. The subjects of the experiment were ten Silver King pigeons.

In Fig. 4.4 the results of this experiment are plotted in four sets of curves, one set for each combination of levels of P1 and delay. Each set of curves plots matching accuracy against length of P2, with ISI as the parameter, and each point on the curves represents 240 trials. Both P1 and P2 were significant effects, yielding increasingly higher performance with increased length of presentation, as would be expected from the previous studies of presentation time. As

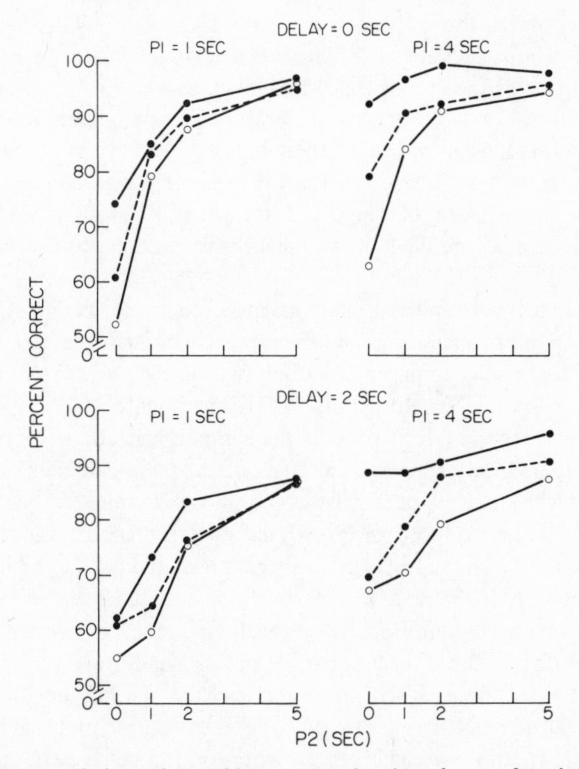

FIG. 4.4 Accuracy of delayed matching to sample plotted as a function of P2 length following an initial P1 of 1 or 4 sec and an ISI of (●——●) 0; (●– – –●), 2; or (○——○), 5 sec; top curves plot data for 0-sec delay and bottom curves plot data for 2-sec delay. (From Roberts & Grant, 1974.)

expected, the 2-sec delay led to a lower level of performance than the 0-sec delay. With regard to the effect of spacing, performance became progressively worse the longer the ISI, and this effect was highly significant statistically. It appears that introducing a time interval between sample presentations lowers performance, a result that conflicts with the human data and appears to support the trace strength and decay model.

To formalize our theory somewhat, we devised an elementary mathematical model and determined its predictions for this experiment. It was assumed that an initial 1 sec of presentation time would establish a starting level of 30 units of trace strength and that every second thereafter would add 50% of the difference between the current level and an upper limit of 100 units. Furthermore, it was assumed that each second of delay would decrease trace strength by 30% of the current level. The parameters of the experiment reported were then used to generate the theoretical curves seen in Fig. 4.5. These curves are plotted for the 0-sec delay condition only, the 2-sec delay curves showing a similar pattern at a lower level of trace strength. Examination of the curves shows that they are in good agreement with the empirical curves shown in Figure 4.4 when the delay is set at 0 sec. Unfortunately, the data at a delay of 2 sec do not match the theoretical curves as nicely. At the 0-sec delay, several of the important aspects of the theoretical curves are seen in the empirical curves, in addition to the main effects of the four variables. One of these is the greater effect of ISI after a P1 of 4 sec than after a P1 of 1 sec, and the other is the increased rates of growth of the curves as a function of length of P2 when P1 is short and ISI is long. Statistical analysis of the data showed significant interactions of P1 × ISI, P1 × P2, and P2 × ISI.

The experiment just reported examined the effects of spacing a single repetition on DMTS performance. In another experiment (Roberts, 1972b, Experiment III), pigeons were exposed to either two or eight successive presentations of a sample stimulus. In this study, the experimenter did not have precise control over the presentation time of the sample; instead, a colored field (red, yellow, green, or blue) appeared on the center key and remained on until the subject pecked the key. A peck on the center key terminated one presentation of the sample and introduced an ISI, which was terminated by another presentation of the sample. The sample stimulus was presented in this fashion either two times (2P) or eight (8P), and ISIs of 0, 1, 3, 6, and 10 sec were interpolated between successive presentations under each frequency of presentation. In this procedure, an ISI of 0 sec meant that 2P and 8P conditions were the same as an FR 2 or FR 8 schedule. When an animal pecked the key on the final sample presentation of a trial, a delay of 0, 2, or 8 sec occurred and was followed by a retention test. The three variables of presentation frequency, ISI, and delay were combined in a factorial design, with presentation frequency and delay varied within daily sessions and ISI varied between sessions.

The trace strength and decay theory already outlined predicts that the effect of ISI should be stronger the greater the number of presentations of the sample

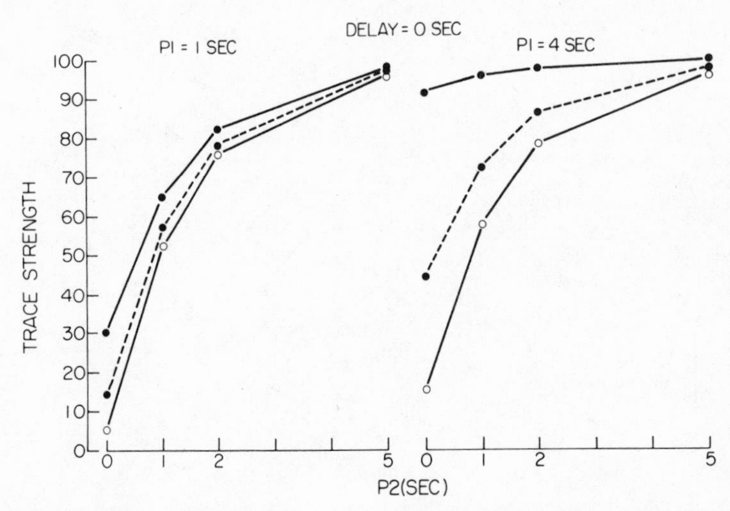

FIG. 4.5 Theoretical data based on a mathematical model in which the first second of presentation time generates 30 units of trace strength, the growth rate parameter is set at .5, and the decay rate parameter is set at .3. (●——●), ISI = 0 sec; (●– – –●), ISI = 2 sec; (○——○), ISI = 5 sec. (From Roberts & Grant, 1974.)

stimulus. The model used to generate the theoretical curves in Figure 4.5 also can be applied to this experiment. It was assumed that each presentation or peck on the sample key increments trace strength by 50% of the difference between its current level and an upper level of 100 units and that each second of delay decreases trace strength by 30% of its current level. Given the further assumption that an initial presentation develops 30 units of trace strength, we used the parameters of the present experiment to calculate the trace strength at the end of the last presentation of a trial for each ISI length at 2P and 8P. The trace strength values revealed by this exercise at ISIs of 0, 1, 3, 6, and 10 sec were 65.0, 60.5, 55.2, 51.8, and 50.4, respectively, at 2P, and 99.5, 76.9, 60.4, 53.2, and 50.7, respectively, at 8P.

In Fig. 4.6, the results of this experiment are shown for ten Silver King pigeons, with each point based on 480 trials. Sets of curves are presented separately for 2P and 8P, each set presenting accuracy as a function of ISI with delay (D) as the parameter. The expected effect of frequency is seen clearly in the figure, performance being considerably higher after 8P than after 2P, and matching accuracy generally declines as the ISI is lengthened. Furthermore, the predicted interaction of frequency and ISI is present, as shown by the fact that correct responses dropped faster as a function of increasing length of ISI at 8P than at 2P. The interaction of ISI × Number of Presentations was statistically significant. A further prediction from trace strength theory was therefore supported.

One final experiment done on spacing concerns the commutativity of ISIs (Roberts, 1972b, Experiment IV). In this experiment, pigeons were tested on

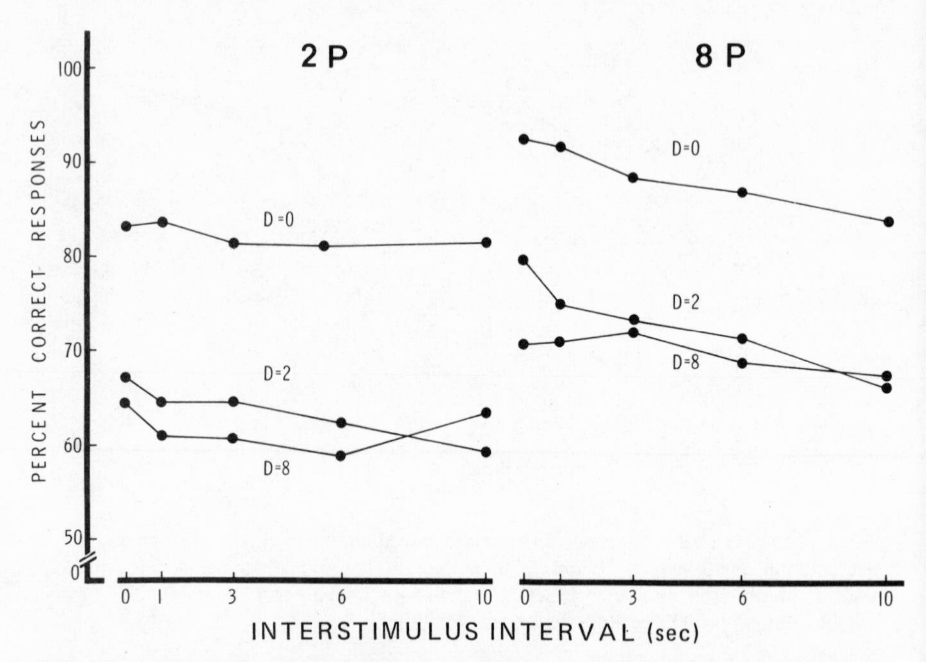

FIG. 4.6 Accuracy of retention in delayed matching to sample plotted as a function of ISI with delay (D) as the parameter. The sample stimulus was presented either twice (2P) or eight times (8P). (From Roberts, 1972b; Copyright 1972 by the American Psychological Association. Reprinted by permission.)

only two presentation conditions, using the same procedures as outlined for the preceding experiment. In both presentation conditions, the sample was presented nine times, with eight ISIs between presentations. In one condition (spaced–massed), the first four ISIs were 10-sec long, and the last four were 0-sec long; just the reverse was the case in the other condition (massed–spaced), the first four ISIs being 0 sec and the last four being 10 sec. Retention was tested at delays of 0, 2, and 8 sec. Because the total spacing time was 40 sec in both conditions, we would expect equally good performance in both conditions if ISIs are commutative in their effects on pigeon STM. Evidence for commutativity of spacing intervals has been reported by Bjork and Abramowitz (1968) for an STM experiment with human subjects. However, the trace strength theory used to account for the results of the preceding experiments predicts a difference between these presentation conditions in favor of the spaced–massed condition. Essentially, the theory predicts the initial gains established in the massed–spaced condition to disappear rapidly with repeated long ISIs at the end but the massed repetitions at the end of a spaced–massed trial to build up a substantially higher level of trace strength.

The outcome of the experiment showed clear evidence of noncommutativity, with a difference between the two conditions of presentation that agreed with the prediction from trace strength theory. When performance was averaged over the three delays, overall performance in the spaced–massed condition was 82.6% correct responses, whereas that in the massed–spaced condition was 75.0%. The difference was highly significant.

All of the experiments done with the pigeon so far support the contention that spaced repetition leads to a lower level of performance on DMTS as compared to massed repetition and that the loss in performance resulting from spacing is related directly to the length of the spacing interval. This result is at direct variance with the findings of human STM studies, in which increased spacing increases level of retention. Furthermore, all of the manipulations involving spacing carried out in the experiments reported have yielded results that are predicted from a theory holding that trace strength grows in the presence of a sample stimulus and decays in its absence. These findings suggest that human and pigeon experiments may be testing memory systems that operate according to quite different principles.

It can be argued that the use of verbal items, which involve language as an encoding device in human experiments, is a critical factor in the differential effect of spacing. However, there is some evidence now that the facilitative effect of spacing appears in nonhuman mammals. Robbins and Bush (1973) tested great apes on a series of two-choice discrimination problems with Trials 1 and 2 on a given problem separated by different numbers of interpolated trials on other problems. When the apes were tested for retention on a third test trial, it was found that memory at longer retention intervals improved as the spacing interval between Trials 1 and 2 increased. Another primate study of DMTS carried out by Medin (1974) found a facilitative effect of spacing in pigtailed monkeys. Medin's study showed that spacing aided retention only under conditions that minimized PI effects. That is, the spacing effect appeared when different stimuli were used on each trial of a daily session, but no effect of spacing was noted when the same stimuli were used on all trials. In an experiment with rat subjects, Roberts (1974) used a delayed alternation test of short-term retention and varied the spacing interval between initial forced runs to one side of a T maze. When rats were allowed to make a free choice as a test of retention, it was found that at long delays they remembered the side last entered better if the initial forced runs had been spaced apart in time than if they had been massed. As with the data discussed in the preceding section, it is tempting to speculate that comparative differences in memory mechanisms underlie these differences in performance. We have entertained the possibility that the spacing effect is characteristic of mammalian memory but is totally missing in avian memory and that this difference may reflect a fundamental species difference in the neural mechanisms responsible for processing and retaining information. A good deal more research is needed, however, to test the validity of this hypothesis. The apparatuses in which these experiments were carried out were different, Wisconsin General Test Apparatus for apes, T maze for rats, and Skinner box for pigeons. Furthermore, the types of memory tests were different in each case. Species comparisons of the effects of spacing on retention in comparable situations seem to be called for now. Also, a greater variety of species would need to be studied in order to determine the validity of this hypothesis. The findings made so far certainly suggest that such experiments are a worthwhile venture.

PROACTIVE INTERFERENCE

Studies of proactive interference (PI) involve a control group that learns only B and an experimental group that learns A prior to B. Both groups are then tested on B. If test performance on B is poorer in the experimental group than in the control group, then the prior learning of A has interfered with performance on B. When the test on B is delayed following the acquisition of B, the primary question of interest is whether or not the prior learning of A has affected the retention of B. That is, did the prior learning of A cause B to be forgotten more rapidly than would have been the case had A not been learned? To determine whether A increased the rate of forgetting of B, it is necessary to employ both an immediate and a delayed retention test. If an equal amount of interference is obtained on both tests, then the interference has not resulted from a retention deficit attributable to prior learning. Instead, the interference resulted from a general lowering of performance on B produced by the prior presentation of A. If interference is obtained on the delayed test and not on the immediate test, however, then the interference has resulted from a retention deficit attributable to prior learning. A third possibility is that interference is obtained on both the immediate and delayed tests, but with more interference on the delayed test. In this case, the interference effect has resulted from a combination of a general lowering of performance on B produced by A and of a retention deficit on B produced by A.

Several studies of PI in human short-term recall memory have obtained evidence of a retention deficit attributable to prior learning (e.g., Keppel & Underwood, 1962; Loess, 1964). In these studies, a trigram composed of three consonants is presented, and then, following a delay of several seconds in which the subject counts backward, the subject is tested for recall of the trigram. Performance at short retention intervals (for example, 3 sec) is unaffected by the number of prior trigrams which the subject has learned. However, at longer retention intervals (for example, 18 sec) performance decreases as the number of prior trigrams the subject has learned increases from zero to four. The learning of prior trigrams therefore increases the rate of forgetting of the current trigram. Hawkins, Pardo, and Cox (1972) have found that a retention deficit attributable to prior learning is obtained also when a recognition, rather than a recall, paradigm is employed.

Studies of PI in animal long-term memory (LTM) also have revealed evidence of a retention deficit attributable to prior learning. Investigations of PI in rat LTM typically involve first training the animals on passive avoidance (no-go), followed by training on active avoidance (go). The typical finding in this situation (Spear, 1971) is that the go response dominates at short delays and becomes progressively less dominant as the delay increases. In addition, the interference condition and the control condition (learned only the go response) are equivalent in retention of the go response on an immediate test. However, with the passage of time, the interference condition shows a more rapid rate of forgetting of the go response than does the control condition. Therefore, the prior acquisition of the no-go response results in a retention deficit. Burr and Thomas (1972) have obtained evidence of a retention deficit attributable to prior learning in pigeon LTM. In the Burr and Thomas study, experimental birds were trained on a wavelength discrimination in Stage 1, followed by a reversal of the Stage 1 discrimination in Stage 2. Control birds learned only the Stage 2 discrimination. All birds were then tested for generalization along the wavelength dimension either immediately or 24 hr after reaching criterion on the stage two task. The immediate-test gradients of experimental and control birds did not differ, but in the delayed test the experimental gradient was flatter and showed less area shift than the control gradient.

On the basis of these studies, it seems reasonable that one variable that may control short-term forgetting in the pigeon is the presence or absence of prior learning. To investigate this possibility we conducted a study to investigate PI in pigeon STM (Grant & Roberts, 1973, Experiment 1). The DMTS procedure was used and two conditions were employed. In the experimental condition, two colored-field sample stimuli, S1 and S2, were presented in immediate succession on the center key. Following a delay after the termination of S2, S1 and S2 were presented simultaneously for a choice, with choice of S2 reinforced. In the control condition, the procedure was identical except that only one sample stimulus, S2, was presented at input. In the first experiment, an FR 1 was required on S1 to introduce S2 in the experimental condition and an FR 1 was required on S2 to introduce the delay, either 0, 1, or 4 sec, in both conditions. Eight White King pigeons served as subjects. Within each daily session of trials, each bird was tested repeatedly under all combinations of the conditions and delay variable.

The results of this experiment are shown in Fig. 4.7, in which retention curves for the experimental and control conditions are presented with each point based on 3,072 trials. Delay and conditions strongly influenced performance on the retention tests; both retention curves show the typical negatively accelerated decline in matching accuracy as the delay is lengthened. The control condition was superior to the experimental condition at each delay and so indicated that the prior presentation of S1 interfered with choice of S2 at the time of test. However, there was no suggestion that S2 was forgotten more rapidly in the

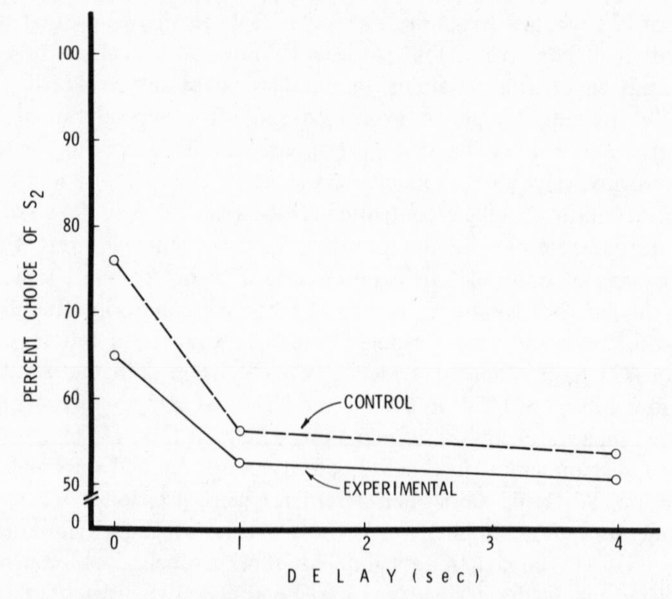

FIG. 4.7 Retention curves for the experimental (interference) and control conditions. (From Grant & Roberts, 1973; Copyright 1973 by the American Psychological Association. Reprinted by permission.)

experimental condition than in the control condition. In fact, the control curve showed slightly faster forgetting than the experimental curve. Therefore, the interference obtained did not result from a retention deficit produced by the prior presentation of S1. Instead, the prior presentation of S1 produced a general lowering of performance.

To account for these findings, a model was proposed based on the competition of independent memory traces. The model extends trace strength theory (Roberts, 1972b; Roberts & Grant, 1974) to the interference situation. It holds that successive stimulus events form separate and independent memory traces. Whereas both traces decay with the passage of time, the presence of one trace in the memory system does not affect either the initial strength or the rate of decay of a second trace. Interference effects are held to be the product of competition between traces at the time of test, with the stronger trace determining the choice responses. The degree of interference is dependent on the extent to which the trace strength distribution of the interfering memory (i.e., S1) overlaps or exceeds the trace strength distribution of the test memory (i.e., S2).

Several predictions derived from the independence and competition model were supported by the data of our initial experiment. First, because at the time of test the trace of S1 had longer to decay than the trace of S2, the trace of S2 should be stronger on the majority of trials. Therefore, the model predicts that S2 should be chosen on the majority of trials (greater than 50%) in the experimental condition. This prediction was confirmed generally, although only performance at the 0-sec delay was found to be significantly above chance in the experimental condition. Second, the model predicts that S1 should interfere with choice of S2 at the time of test. This prediction was confirmed by the significant difference between the conditions. Third, because the model holds that the trace of S1 does not affect the rate of decay of the trace of S2, the model predicts the curves should fall in parallel as the retention interval increases. The generally parallel course of the experimental and control curves verified this prediction.

To test the independence and competition model of interference effects in pigeon STM further, a second experiment was performed (Grant & Roberts, 1973, Experiment II). As the independence and competition model holds that the degree of interference is related directly to the degree of overlap between the trace strength distributions of S1 and S2, it should be possible to alter the amount of interference by manipulating the trace strength of the two conflicting memories. Techniques for manipulating trace strength suggested by Roberts' (1972b) studies of repetition and spacing were employed to manipulate independently the trace strengths of S1 and S2. It was predicted that repetition of S1 at input would increase the amount of interference at the time of test, whereas repetition of S2 at input would decrease the amount of interference at the time of test. The introduction of an interstimulus interval (ISI) between the termination of S1 and the onset of S2 at input should decrease the amount of interference at the time of test. The design of the experiment involved the factorial manipulation of four variables; two levels of S1 repetition (FR 1 or FR 30), two levels of S2 repetition (FR 1 or FR 5), two levels of ISI (0 or 10 sec), and two levels of delay (0 or 3 sec).

The percentage of S2 choices for each of the 16 experimental conditions and each of the four control conditions are shown in Table 4.1. Collapsing across delays, significant interference was obtained in all four experimental conditions when the ISI was set at 0 seconds. In contrast, none of the four experimental conditions differed from the control when the ISI was set at 10 sec. As predicted, interference from S1 was clearly present at a short ISI and minimal at a long ISI. Because no interference was obtained when the ISI was set at 10 sec, the 0-sec ISI data were used to test the prediction that the degree of interference would be increased by raising the FR on S1 and lowered by raising the FR on S2. When the 0-sec ISI percentages are averaged over the two values of the FR on S2 variable, the difference between the FR 30 on S1 mean and the control mean is 9.6 percentage points, whereas the difference between the FR 1 on S1 mean and the control mean is 5.1 percentage points. A test between these differences showed that interference under FR 30 was significantly greater than the interference under FR 1. A similar analysis, in which percentages were averaged over the FR on S1 variable, showed the difference between the FR 1 on S2 mean and its control mean to be 8.9 and the difference between the FR 5

TABLE 4.1
Percentage of S2 Choices under Each Condition[a]

	FR1[b]		FR30[b]		
Delay (sec)	0-sec ISI	10-sec ISI	0-sec ISI	10-sec ISI	Control
			FR1[c,d]		
0	60.5*	68.2*	56.9*	66.5*	74.5*
3	53.8	48.8	49.1	56.3*	53.5
$\bar{X}$	57.2	58.5	53.0	61.4	64.0
			FR 5[c,d]		
0	75.2*	77.8*	67.4*	76.4*	76.7*
3	54.9	59.0*	52.9	55.7*	59.9*
$\bar{X}$	65.1	68.4	60.2	66.1	68.3

[a]From Grant and Roberts (1973); Copyright 1973 by the American Psychological Association. Reprinted by permission.
[b]Number of pecks required on S1 to introduce S2.
[c]Number of pecks required on S2 to introduce the delay.
[d]Data marked with an * are significantly above chance at the .05 level.

on S2 mean and its control mean to be 5.8. Although this difference was in the predicted direction, it did not reach significance. In general, the data from this experiment support all of the predictions derived from the independence and competition model of interference effects.

The data in Table 4.1 also provide additional evidence against the notion that the presentation of S1 increases the rate at which S2 is forgotten. It can be argued that the data in Fig. 4.7 do not rule out the possibility that S1 increases the rate at which S2 is forgotten, for this effect may be masked by a floor artifact at the 1- and 4-sec delays in the experimental condition. As shown in Table 4.1, four of the eight experimental means were above chance at the 3-sec delay (those marked by an asterisk). For each of these conditions, retention loss scores were calculated by subtracting the 3-sec percentage of S2 choices from the 0-sec percentage. The means of these retention loss scores were then compared to the mean retention loss score for the appropriate control condition. In none of these tests was the retention loss for the experimental condition found to be significantly greater than that for the control condition. These data provide additional evidence against the notion that the prior presentation of S1 increases the rate of forgetting of S2. The data support the hypothesis that the prior presentation of S1 produces a general lowering of performance across all delays.

Although the data so far presented have been interpreted as supporting an independence and competition model, an alternate interpretation in terms of a theory based on trace strength limitation is possible. According to this theory, there is a limit on the total amount of trace strength that can be stored in memory. Furthermore, it may be assumed that storing the trace of S1 occupies a sufficient proportion of the capacity of the memory system to result in an incomplete storage of the S2 trace in the experimental condition. In the control condition, no such incomplete storage of the S2 trace would occur because S1 was not presented. As a result, the trace of S2 may be stronger in the control condition than in the experimental condition and lead to a decreased percentage of S2 choices in the experimental condition. The trace strength limitation theory can account also for the effects of the repetition and spacing variables on the interference effect.

To test the trace strength limitation theory we conducted an experiment (Grant & Roberts, 1973, Experiment III) in which experimental trials consisted of the successive presentation of S1 and S2. However, retention of S2 was tested by presenting S2 simultaneously with a new stimulus, S3. The limitation theory and trace competition theory make different predictions concerning the way in which this procedure should affect choice of S2 relative to a control condition in which S1 is not presented. Limitation theory predicts that interference should be found, because the presentation of S1 should limit the strength of the S2 trace regardless of the stimuli presented at the time of test. Trace competition theory predicts no interference effect because no stimulus corresponding to the trace of S1 is present at the time of test. In order to make a strong test of limitation theory, S1 was presented at two levels of repetition, FR 1 and FR 30, and S2 was presented only once, FR 1, in both conditions. The experiment revealed no differences between the experimental conditions, FR 1 and FR 30 on S1, and a control condition, which presented only S2. If the percentage of S2 choices are averaged over the three delays used (0, 1, and 3 sec) for each condition, the mean percentages of correct choices for experimental conditions FR 30 and FR 1 and for the control condition are 67.4, 67.4, and 67.3, respectively. Therefore, these data clearly rule out trace strength limitation theory as an adequate account of interference effects in pigeon STM.

According to the trace competition model, the prior presentation of S1 interferes with choice of S2 only to the extent that the strength of the S1 trace is equal to or greater than the strength of the S2 trace at the time of test on some proportion of trials. Therefore, if the trace strength of S1 and S2 are equal at the time of test, the model predicts a 50–50 split in choice of S1 and S2 at the time of test. As the trace of S1 is made progressively stronger than the trace of S2 at the time of test, the percentage of S2 choices should drop below 50%. As the trace of S1 is made progressively weaker than the trace of S2 at the time of test, the percentage of S2 choices should rise above 50%.

These predictions were tested in an experiment conducted by Roberts and Grant (1974, Experiment 3) in which the timer-controlled DMTS procedure was employed. In the experiment depicted in Fig. 4.4 (Roberts & Grant, 1974, Experiment 2), it was found that a stimulus presented for 4 sec and followed by a delay of 2 sec yielded the same percentage of correct responses on DMTS trials as a stimulus presented for 2 sec and followed by an immediate retention test, 85%. Because the level of performance was equal, it was hypothesized that the level of trace strength at the time of test was equivalent in the two conditions. When these observations are extended to the interference situation, presenting S1 for 4 sec and following this immediately by a 2-sec presentation of S2 leads to the prediction that a 50% level of choice of S2 is to be found on an immediate test. Furthermore, as S1 is varied systematically and S2 is held constant at 2 sec, choice of S2 should become greater than 50% as S1 is made shorter than 4 sec and should become less than 50% as S1 is made longer than 4 sec. These predictions were tested in the first block of testing, and three additional blocks of testing were also conducted. In these further tests, the values of S1, ISI, and S2 were changed from those of the first block. Each block of days, in which the parameters were held constant across days, was designated as a separate experiment, being labeled Experiments A, B, C, and D. Ten Silver King pigeons served as subjects and each block represented six sessions of testing. Each daily session consisted of 48 trials for each bird.

The percentages of S2 choices for Experiments A, B, C, and D are shown in Table 4.2. First, in Experiment A, the prediction that choice of S2 would approximate 50% when S1 was 4 sec and S2 was 2 sec was not supported. In fact, choice of S2 under these conditions was markedly higher than 50%, being 69.2%. Even more surprising was the finding that choice of S2 remained significantly above 50% when S1 was lengthened to 5 and 6 sec. In Experiments B, C, and D, attempts were made to produce a level of S2 choice below 50% by operations designed to weaken the trace of S2 and strengthen the trace of S1. As can be seen in the table, none of these manipulations was successful in producing a level of S2 choice significantly below 50%. Only in Experiment 3B did choice of S2 drop below 50%, at S1 lengths of 5 and 6 sec, but these percentages were not significantly different from chance.

These data replicated our previous findings (Grant & Roberts, 1973) that presenting S1 prior to S2 at input interfered with choice of S2 at the time of test. In addition, the amount of interference increased as S1 was lengthened and as S2 was shortened. These findings indicate that competition on the basis of strength between independent memory traces is a factor in interference effects in pigeon STM. However, some factor other than competition must be added to the theory to explain the persistent preference for S2. We proposed that this additional factor might be an innate tendency to respond on the basis of the more recent of a series of sequential events. This tendency was assumed to be innate primarily because choice of S2 did not become progressively greater in the experimental condition, relative to the control condition, as a function of sessions either in this experiment or in previous ones (Grant & Roberts, 1973). An interaction between conditions and practice would be expected if the recency factor was based on learning.

TABLE 4.2
Percentage of S2 Choices in Experiments A, B, C, and D[a]

	Length of S1 (sec)[b]								
Experiment	0	1	2	3	4	5	6	7	9
A (ISI = 0 sec, S2 = 2 sec)	90.4*		80.4*	69.2*	69.2*	60.8*	65.8*		
B (ISI = 1 sec, S2 = 1 sec)	75.8*		62.5*	61.7*	55.4	48.8	46.7		
C (ISI = 1 sec, S2 = 1 sec)	76.7*	67.9*		62.5*		62.5*		60.8*	54.2
D (ISI = 0 sec, S2 = 1 sec)	75.4*	68.8*		69.6*		62.9*		57.1	53.3

[a]From Roberts and Grant (1974).
[b]Asterisk indicates significantly different from 50% at the .05 level by *t* test.

On the basis of the findings of these studies of PI in pigeon STM it may be argued that the processes that control short-term forgetting in the pigeon are unaffected by prior input. Acceptance of this conclusion may be premature, however, because the paradigm employed to study PI in pigeon STM differs from the paradigm employed in most PI studies that have obtained a retention deficit attributable to prior learning. One of the major differences involves the nature of the interfering event. In the pigeon STM paradigm, an intratrial interference paradigm, the interfering event consists of a stimulus presented on the same trial as the stimulus to be remembered, whereas in the human STM and in the pigeon and rat LTM interference paradigms, the interfering event consists of one or more complete trials of learning.

This paradigm difference was eliminated by employing an intertrial interference paradigm to investigate PI in pigeon STM (Grant, 1975b). The intertrial paradigm eliminated the paradigm difference discussed above by employing a complete DMTS trial as the interfering event. In the interference condition, two DMTS trials, T1 and T2, were presented in immediate succession; correct and incorrect colors on T1 being reversed on T2. In the control condition, T1 was not presented and T2 occurred in isolation. It seems reasonable to predict that T1 would interfere with performance on T2; the critical question is whether T1 produces a general lowering of performance on T2 (as does S1 in the intratrial paradigm) or whether T1 increases the rate of forgetting on T2 (and thereby produce a retention deficit attributable to prior learning).

In the first experiment (Grant, 1975b, Experiment I), the effect of the prior presentation of T1 on T2 retention was assessed across T2 delays of 0, 2, 6, and 10 sec. In the interference condition, T1 was presented either one, four, or six times before T2, labeled 1P, 4P, and 6P conditions, respectively. Increasing the degree of learning of the interfering event increases the amount of interference in both human STM (Loess, 1964; Wickens, 1970) and rat LTM (Spear, Gordon, & Chiszar, 1972). As previously discussed, increasing the amount of time during which the bird was exposed to the interfering stimulus increased the magnitude of the interference effect in the intratrial paradigm (Grant & Roberts, 1973; Roberts & Grant, 1974).

In the experiment, ten Silver King pigeons were tested repeatedly on all combinations of the conditions and delay variables. Each bird was tested on 16 trials per day for 14 days. Retention curves for the 1P, 4P, 6P, and control conditions are shown in Fig. 4.8, each point on the curves representing 240 trials. An interference effect was obtained in each of the interference conditions because the control was superior to each of the interference conditions at each delay, with the exception of the equality of the control and 1P conditions at the 10-sec delay. The 1P and 6P conditions showed similar effects of T1 on T2 retention. In each of these conditions, the rate of T2 forgetting was greater over the first 6 sec than in the control condition. Both conditions demonstrated

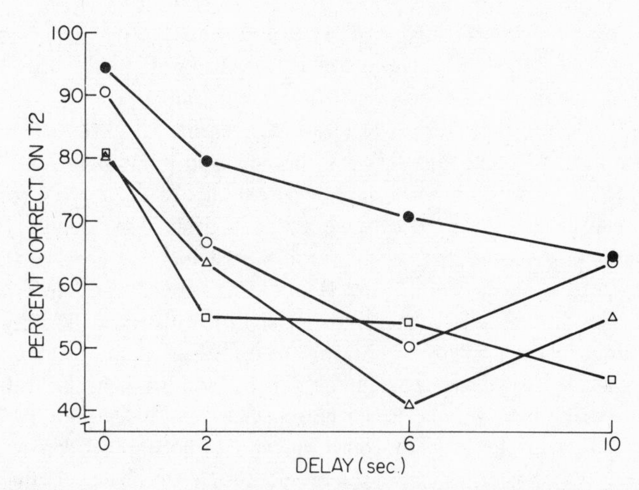

FIG. 4.8 Retention curves for delayed matching to sample trials preceded by either (○), one; (□), four; or (△), six interfering trials or by no interfering trials (control, ●). (From Grant, 1975b; Copyright 1975 by the American Psychological Association. Reprinted by permission.)

better T2 retention at the 10-sec than at the 6-sec delays. The 4P curve paralleled the 1P and 6P curves for the first 2 sec of the retention interval, and then, unlike the 1P and 6P conditions, the 4P condition showed almost no T2 forgetting over the next 4 sec. Also, the level of T2 retention in condition 4P was lower at the 10-sec than at the 6-sec delay; this finding did not follow the pattern of the other two interference conditions. A comparison of the 1P and 6P conditions shows that increasing the number of T1 presentations lowers T2 retention and thereby increases the interference effect by a relatively constant amount across T2 delays.

The type of interference effect obtained in the intertrial paradigm is in marked contrast to the type of interference effect obtained in the intratrial paradigm (compare Figs. 4.7 and 4.8). In the intratrial paradigm, the prior presentation of S1 produces a general lowering of performance but does not affect the rate of S2 forgetting. In the intertrial paradigm, however, the prior presentation of T1 increases the rate of forgetting on T2. Under certain conditions, therefore, a retention deficit attributable to prior learning can be obtained in pigeon STM.

Grant (1975b, Experiment III) conducted another experiment that attempted to determine the critical difference between these paradigms, which is responsible for producing the different types of interference effects. The T1 trial was dissected into components, and these various components then were presented as the interfering event; four such interfering events were employed. In Condition I, a complete DMTS trial was used as the interfering event (intertrial paradigm). In Condition II, the T1 sample stimulus plus only the correct side key was used as the interfering event. In Condition III, only the T1 correct side key served as the interfering event and in Condition IV, only the T1 sample stimulus served as the interfering event (intratrial paradigm).

In Fig. 4.9, retention curves for the four interference conditions (I, II, III, and IV) and the control condition (V) are shown, each point based on 360 trials. Each interference condition had a higher percentage of correct T2 responses at the 0-sec delay and a lower percentage at the 6- and 10-sec delays than in the control condition. At the 2-sec delay, conditions I and III were slightly superior to the control, whereas Conditions II and IV were slightly inferior to the control. In Condition I, T2 was forgotten significantly more rapidly over the first 6 sec than in the control condition. In Conditions II and IV, the rate of T2 forgetting was slightly faster than in the control condition over the first 2 sec, although the difference did not approach significance. The rates of T2 forgetting were about equal in the control condition and Condition III across the entire retention interval. Statistical analyses revealed that Condition I, the intertrial condition, was the only condition that produced any reliable interference, with a retention deficit attributable to prior learning present over the first 6 sec of the retention interval.

Unfortunately this experiment did not answer the question of whether the intertrial and intratrial interference paradigms differed qualitatively with respect to the type of interference effect produced. That is, given that an interference effect is produced, does the intertrial paradigm always produce a retention deficit and the intratrial paradigm always produce a general lowering of performance? Whether or not the paradigms differ qualitatively, they do differ quantitatively. A prior complete trial provides a more potent source of interference than does a prior stimulus occurring on the same trial as the stimulus to be remembered.

Two additional experiments were conducted. In the first of these (Grant, 1975b, Experiment II), an encoding or storage failure interpretation of the retention deficit was tested. It was found that no interference was obtained when neither of the T1 stimuli appeared in a reversed role on T2. A retention deficit was obtained either when the correct and incorrect colors on T1 were reversed on T2 or when only the T1 incorrect color appeared on T2 in the role of the sample stimulus. These data support the hypothesis that the retention deficit in pigeon STM results from the competition of conflicting memories at the time of test. In the second of these experiments (Grant, 1975b, Experiment IV), it was found that the retention deficit in pigeon STM was sensitive to the intertrial interval (ITI) separating the termination of T1 and the onset of T2. At a 2-sec ITI, a large retention deficit attributable to prior learning was obtained. At 20- and 40-sec ITIs, however, negligible interference was obtained. Moreover, there was no evidence to suggest that prior learning increased the rate of forgetting of subsequent learning when the ITI was set at 20 sec or longer.

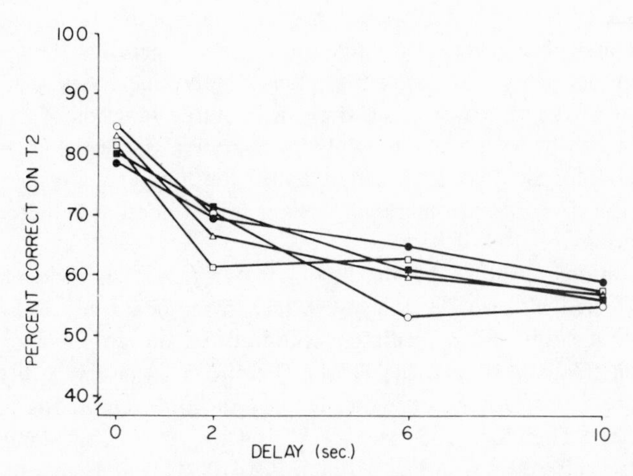

FIG. 4.9 Retention curves for the four interference conditions (○, I; □, II; ■, III; and △, IV) and the control condition (●, V). (From Grant, 1975a. Copyright 1975 by the American Psychological Association. Reprinted by permission.)

The results reported by Grant (1975b) demonstrate that under certain conditions a prior, conflicting memory increases the rate at which a subsequent memory is forgotten in pigeon STM. Although these data are not inconsistent with theoretical models emphasizing the loss and recovery of associations or memory attributes (e.g., Keppel & Underwood, 1962; Spear, 1971) or with models emphasizing temporal discrimination factors (e.g., D'Amato, 1973; Gleitman, 1971), the data are interpretable also in terms of the independence and competition model previously discussed. The interesting aspect of the independence and competition model is that it can account for both the general lowering of performance obtained by Grant and Roberts (1973) and the retention deficit obtained by Grant (1975b). The feature of the model that makes this possible is the assumption that forgetting in pigeon STM is a negatively accelerated function of time since the termination of the stimulus. Therefore, a constant amount of interference is produced across delays when the two conflicting memories begin to decay at approximately the same point on the negatively accelerated decay function. Under these circumstances, both traces decay or lose strength at approximately the same rate. Thus, the degree of overlap between the trace strength distributions remains relatively constant as the retention interval increases, and the degree to which the first trace competes for dominance with the second trace remains constant as the retention interval increases and produces a constant amount of interference across delays.

In contrast, increasing interference with increasing delay is produced when the first trace has undergone a period of trace decay before the second memory begins to decay. Under these circumstances, the first, interfering memory has already undergone a period of rapid trace decay before the second, test memory begins to decay. Therefore, as the retention interval increases beyond 0 sec, the second memory loses trace strength more rapidly than does the first trace, because trace strength loss is a negatively accelerated function of time since the termination of the event. As the retention interval increases, the degree of overlap between the two trace strength distributions increases and results in increasing interference with increasing delay and a retention deficit attributable to prior learning.

The conditions specified by the model that result in the rate of forgetting being unaffected by prior learning were met in the Grant and Roberts (1973) study. In that study, S2 immediately followed S1 and an FR 1 or FR 5 was required on S2 to introduce the delay. Therefore, S1 had very little time to decay before S2 began to decay (that is, only the time required to peck S2 one or five times). Therefore, the trace of S1 and the trace of S2 should decay at approximately the same rate as a function of delay and result in a constant amount of interference across delays and nondifferential rates of forgetting as a function of the presence or absence of prior learning.

The conditions specified by the model that result in the rate of forgetting being increased by prior learning were met in the Grant (1975b) study. In that study, the T1 memory had a minimum of 6 sec of decay time before the T2 sample stimulus memory began to decay (that is, the 2-sec reinforcement or blackout following the side key response on T1 plus the 4-sec presentation of the T2 sample stimulus). The T1 memory had therefore already undergone a period of rapid trace decay before the T2 sample stimulus memory began to decay. Therefore, because of the negatively accelerated decay function, the T2 memory should be forgotten more rapidly than the T1 memory as the retention interval increased. This would result in the means of the two trace strength distributions becoming closer as the retention interval increased and lead to increasing interference as the retention interval increased and differential rates of forgetting as a function of the presence or absence of prior learning. To account for the decrease in the magnitude of interference at the 10-sec delay, it could be assumed that the T1 memory had been lost from STM by this time on the majority of trials.

One major question regarding PI effects in pigeon STM is whether or not the type of interference effect obtained in the intertrial paradigm differs qualitatively from the type of interference effect obtained in the intratrial paradigm. That is, does the memory of a prior complete trial interfere with a later memory in a fundamentally different way from the way in which the memory of a prior stimulus interferes with a later memory? The independence and competition model holds that whether or not prior learning affects the rate of forgetting of subsequent learning is a function of the amount of decay of the interfering memory that occurs before the test memory begins to decay, not of the nature of the interfering event. If the interfering memory decays very little before the test memory begins to decay, prior learning does not affect the rate of forgetting of subsequent learning. If the interfering memory has undergone a period of trace decay before the test memory begins to decay, then prior learning does increase the rate of forgetting of subsequent learning. According to the model, therefore, it should be possible to produce both differential and nondifferential rates of forgetting as a function of the presence or absence of prior learning in both paradigms.

To test the above prediction, a study designed to produce both differential and nondifferential rates of forgetting as a function of the presence or absence of prior learning was conducted employing the intratrial interference paradigm (Grant, 1975c). Four Silver King pigeons, with considerable prior matching experience, served as subjects. The birds were also sophisticated with respect to intertrial interference because they had served as subjects in the Grant (1975b) study. The model predicts that if S1 is allowed a period of trace decay before S2 begins to decay, then S1 should increase the rate of forgetting of S2 relative to a control in which only S2 is presented. These conditions were met by presenting S2 for 4 sec at input, as this operation resulted in a minimum of 4 sec of S1 decay time before S2 began to decay. The model also predicts, however, that if S1 is allowed to decay very little before S2 begins to decay, then S1 should not affect the rate of forgetting of S2 relative to a control in which only S2 is presented. These conditions were met by presenting S2 for .5 sec at input, a procedure that resulted in only .5 sec of S1 decay time before S2 began to decay. In the condition in which S2 was presented for 4 sec, S1 was also presented for 4 sec and in the condition in which S2 was presented for .5 sec, S1 was also presented for .5 sec. Each subject was tested in 64 trials per session, 32 interference trials and 32 control trials. The exposure duration of S1 and S2 (either .5 or 4 sec) was manipulated between sessions. The experiment was run for 16 sessions, eight .5-sec presentation sessions and eight 4-sec presentation sessions.

The results of this experiment are shown in Table 4.3. As can be seen, the magnitude of interference, in percentage terms, across delays of 0, 2, 5, and 10 sec was 5.9, 15.2, 21.9, and 9.4 when input stimuli were presented for 4 sec. When input stimuli were presented for .5 sec, however, the magnitude of interference across the same delays was 10.6, 5.4, 8.6, and 4.7. As predicted by the model, therefore, prior learning increased the rate of forgetting of subsequent learning when the interfering memory had undergone a period of trace decay before the test memory began to decay (that is, in the 4-sec presentation condition). In contrast, prior learning did not affect the rate of forgetting of subsequent learning when the interfering memory had relatively little time to decay before the test memory began to decay (that is, in the .5-sec condition). These data indicate that the nature of the interfering event is not the critical factor determining the type of interference effect produced. Instead, the data

TABLE 4.3
Percentage of S2 Choices in the Control and Experimental Conditions and the Magnitude of the Interference Effect as a Function of Presentation Time and Delay

Condition	Delay (sec) 0	2	5	10
S1 and S2 each presented for 4 sec				
Control (S2 only)	99.2	92.1	89.1	81.7
Experimental (S1 and S2)	93.3	76.9	67.2	72.3
Magnitude of interference (control – experimental)	5.9	15.2	21.9	9.4
S1 and S2 each presented for .5 sec				
Control (S2 only)	76.2	62.5	63.7	64.1
Experimental (S1 and S2)	65.6	57.1	55.1	59.4
Magnitude of interference (control – experimental)	10.6	5.4	8.6	4.7

support the notion that the critical factor determining the type of interference effect produced is the amount of memory decay of the first trace that occurs before the second trace begins to decay. These data are clearly in line with predictions derived from the independence and competition model of interference effects in pigeon STM.

SUMMARY AND CONCLUSIONS

We have reported a number of experiments in which we have tried to discover how certain variables affect pigeon STM in a DMTS task. The effects of two major variables, length of exposure to the sample stimulus and length of time in the absence of the sample stimulus, seem to be quite straightforward. The accuracy of matching improves as exposure duration increases, and matching deteriorates progressively in the absence of the sample stimulus, whether the removal of the sample is introduced as a delay between sample and test or as an ISI between successive presentations of the sample. The curves that depict the relationship between performance and either presentation time or delay have a negatively accelerated shape. In still further experiments, we complicated things somewhat by introducing conflicting information prior to presentation of the sample stimulus. The intent was to study possible interference effects in STM by using a proaction design. It was found that presenting conflicting information prior to the sample presentation produced both general interference at all delays and increasing forgetting across delays. Although it initially appeared that the type of interference effect obtained depended on the type of interference paradigm used (intratrial vs intertrial), more recent evidence suggested that the critical factor was the length of time between the termination of the interfering stimulus and the termination of the sample stimulus.

On the theoretical level, we have proposed a model of pigeon STM which holds that a memory trace of the sample stimulus grows in strength with exposure time and decays in the absence of the sample. These assumptions largely account for the effects of presentation time, delay, and spacing on performance. In order to deal with interference effects, the added assumption has been made that multiple traces entering the memory system remain independent of one another in strength and rate of decay but compete with one another on retention tests. By taking into account the relative rates of decay of the traces of interfering and sample stimuli, it has been possible to predict both the general interference effect and the increased rate of forgetting effect found in interference experiments. Although the success of this model in accounting for pigeon behavior in STM experiments has been good, it has not been perfect. One inadequacy of the theory has been in accounting for the effects of varying the lengths of presentation of S1 and S2 in an intratrial interference experiment. When the lengths of S1 and S2 were adjusted to provide equal trace strengths at the moment of the retention test, birds still showed a strong preference for S2, in contrast to the chance level of performance predicted by the theory. In this case, it has been necessary to assume that a factor outside the assumptions of the model is operating, that is, that pigeons have an innate predisposition to respond more readily to the more recently occurring information in a serial presentation.

An attempt has been made to draw some comparative implications from our research where possible. Some of our findings contrast markedly with those reported in studies of human and nonhuman primate memory. One example is the fact that monkeys which have been highly trained on DMTS need only a brief exposure to a sample stimulus in order to encode it for relatively long-term retention, whereas the pigeon's performance on retention tests remains strongly tied to presentation time, even in highly experienced birds. Another difference lies in the finding that monkeys apparently form relatively permanent memories of events occurring on DMTS trials, memories that can interfere with performance on subsequent trials. Sequential interference effects of this sort are either absent or very brief in duration in our pigeon experiments. Finally, the most striking comparative difference noted has been the effect of spaced repetition of a stimulus to be remembered on memory in pigeon and human subjects. Spacing strongly enhances retention in human experiments and just as strongly detracts from performance in pigeon experiments. That birds may differ from mammals in general in the effects of spacing is suggested by some recent findings that apes and rats show improved retention when repetitions of stimuli to be remembered are spaced temporally. Further research using other species and common memory tasks is needed to establish the generality or lack of generality of the comparative differences so far observed.

One final consideration is the direction that further research in pigeon memory should take. The theory we have presented suggests that the pigeon's memory for a sample stimulus in a DMTS task is rather short. Yet, we know from studies of discrimination learning and schedules of reinforcement that pigeons learn behaviors that can be retained for hours, days, or even years. How is it that one type of experiment shows such ephemeral retention and others such prolonged retention? Undoubtedly, the answer has to do with repetition of the same reinforcement contingencies from one trial or one moment to the next in experiments that show long-term retention. It may well be that the events of any single trial or reinforcement are short lasting in the way that memory of a sample stimulus is. A major question then concerns the way in which repeated trials or reinforcements lead to the formation of permanent memories. This has been a traditional problem in human memory, and we suggest that it is an important direction for research in pigeon memory.

ACKNOWLEDGMENTS

Support for preparation of this paper was provided by National Research Council of Canada Grant A7894.

5
Animal Models and Memory Models

Douglas L. Medin
The Rockefeller University

RELATIONSHIP BETWEEN HUMAN AND ANIMAL MEMORY RESEARCH

Given the long and generally successful history of using animal models in the life sciences as a tool for understanding *Homo sapiens,* it is only natural to wonder whether animal models can play any role in understanding complex cognitive processes, such as memory. A quick scan of recent literature on memory shows that both animal memory research (e.g., Deutsch, 1969; Bryne, 1970; Pribram & Broadbent, 1970; Honig & James, 1971; Jarvik, 1972; Spear, 1973; Medin & Davis, 1974) and human memory research (Norman, 1970; Tulving & Donaldson, 1972; Melton & Martin, 1972; Anderson & Bower, 1973; Murdock, 1974) are flourishing. Yet a closer look reveals almost no interaction between these two areas.

Although animals have been a common choice in the past for testing learning theories, research on animal memory has been concerned primarily with developing a theory of the function of the brain and has relied on neurophysiological approaches in searching for the neural basis of memory. In contrast, research on human cognitive processes has been more concerned with developing a model for (human) memory without reference to physiology. A common view seems to be that animal physiological work studies structure whereas human research studies function and that this is a convenient division of labor. However, Estes (1973) has argued that a common theoretical framework may be advantageous in guiding research in these areas because the lack of interaction may stem largely from different theoretical conceptions of a learning or memory experiment.

In the absence of such a common framework, we find animal physiological work suffering from the lack of appropriate constructs to interpret structure–

function relationships, behaviorally oriented work with animals aggravated by the (inappropriate) blanket application of constructs deriving from human work to animal paradigms, and human research increasingly subject to criticism for being narrow in scope and narrow in applicability to any nonstandard paradigms and subject populations. Greater interaction between human and animal memory research is needed, if only to avoid confusion. For example, both areas employ the construct short-term memory, but in human studies it is conceived to last at most 10–15 sec (e.g., Peterson & Peterson, 1959), whereas in the animal memory framework short-term memory usually is conceived of as lasting minutes, hours, or up to a day.

Recent summaries and reviews indicate an emerging interest in behavioral approaches to animal memory (Weiskrantz, 1968; Honig & James, 1971; Jarrard, 1971; Medin, 1972a; and Medin & Davis, 1974). They also testify to the awkwardness of relying on concepts borrowed directly from studies of human memory (further discussion of this point appears in Winograd, 1971), an awkwardness that Jarrard and Moise (1971) maintain arises because investigators have thought in terms of paradigms rather than underlying processes as a basis of comparison.

Developing behaviorally based theories of animal memory may serve as an initial step to ameliorate this situation. Our legacy of theories of animal memory is extremely modest; indeed, if constrained to fairly broad theories, I can list only classical interference theory, recently reviewed and assessed by Gleitman (1971), and the stimulus fluctuation model of Estes (1955a, b). The fluctuation model originally has been advanced to account for such phenomena as the effects of trial distribution on the rate of learning in animals but, as we shall presently see, the model may be easily applied to more conventional memory paradigms.

In this chapter, I propose to present Estes' fluctuation model again in order to bring out some of its implications for animal memory, and then to consider modifications in the model in light of the last 20 years of research on animal memory. In the process of aiming toward these two goals, I hope to provide some overview of the relationship between this work and physiological research, on the one hand, and studies using human subjects, on the other.

Research on human learning and memory recently has reflected great interest in coding processes and code structures. One of the more prominent specific theories about coding is Martin's theory of encoding variability (Martin, 1968). Basically this theory assumes (1) that encoding operations on some nominal stimulus produce a subset of the possible functional stimuli and (2) which of these functional stimuli are activated may vary from occasion to occasion. For example, the word "fan" may be encoded at one time as a device for moving air and at another time as an ardent spectator at some activity (for example, a sports fan). Many learning and memory phenomena may be understood in terms of such coding and variability of coding, according to Martin's theory. Because of the conceptual similarity between "encoding variability" and the fluctuation process of the Estes' model, the present analysis of the stimulus fluctuation theory may be especially timely.

First, however, to give a better picture of the spirit with which this work is conducted, I shall consider briefly the questions of why one wants to do behaviorally oriented research on animal memory and how one can proceed in a comparative research framework when differences in performance appear (say, between human and animal subjects).

Why Do Behavioral Research on Memory in Animals?

One set of answers to this question centers around why one does comparative research on cognitive processes at all, without reference to whether the comparison involves different age groups, cultures, or animals. This general question has been discussed extensively elsewhere (Medin & Cole, 1975), and I only suggest here that inherent in a theory is a natural search for variability, which provides new tests for a theory, tests that increase the scope and power of the theories, or tests that betray them as fundamentally inaccurate or incomplete. Species differences, like age or cultural differences, may provide one such potential source of variability and thereby aid theory construction.

There are, however, reasons why one should like to study memory specifically in animals. First, it is often easier to control an animal's experiences, and animals perform readily in tasks that at best are tedious for human subjects. Second, a theory of memory growing out of animal research can avoid a bias toward verbal processing that has been strong in human memory research; for a long time short-term memory processes in humans have been assumed to be mainly acoustic and only in the last few years has there been an interest in, for example, visual short-term memory (e.g., Kroll, Parks, Parkinson, Bieber, & Johnson, 1970). Animal research may help to disentangle those aspects of human memory that depend uniquely on language from those that do not. Finally, with animal subjects one can impose physiological treatments and create the possibility for a positive interaction between behavioral theory and neuropsychology. For example, Spear (1971) used the concept of retrieval failure to interpret experiments on the Kamin effect (Kamin, 1957), which previously had been assumed to be amenable, primarily, to physiological analysis. A theory of animal memory may or may not turn out to be different from a theory of human memory, but either way such information can be of great value.

How Should One Proceed when a Difference in Memory Performance Is Found?

Historically, there has been some tendency to treat comparisons as ends in themselves. For example, some investigations have been revolved around such questions as whether or not monkeys have a better memory than raccoons. Because such processes as memory can be brought to bear in a number of situations, producing a variety of performances, however, it seems more pertinent to ask how memory works in some species rather than asking how much memory animals in that species have. If monkeys do better than raccoons on a memory task such as delayed response, what are they doing better?

Having observed a difference in performance, one usually attempts to establish the basis of this difference and finally verifies the hypothesized explanation by producing conditions that make the differences disappear. This boils down to using a criterion of necessity and sufficiency in evaluating a theoretical explanation; but something more is involved, if only as a byproduct.

This something more can be called replacing proper nouns or nominal variables by independent variables. Age, culture, species, or type may be used as preliminary designators in searching for process diversity, but at some point the meaning in the nominal variables must be deduced if one is to avoid the circularity of explaining differences by invoking these differences again (also see Trzeworski & Teune, 1970).

How might this work? Consider the following observation on height (which incidentally skirts the whole issue of psychological measurement); American girls show a growth spurt at age 12, whereas American boys show a similar spurt at age 15. Assuming this to be an accurate observation, the comparison would take on meaning only when placed in the context of a suitable theory. Let us suppose that we arrive at a theory of growth suggesting that a growth spurt occurs with the onset of puberty. We may then explain the interaction of sex and age on height by proposing (and then checking) the idea that girls reach puberty before boys. The nominal variables of age, culture, and sex would be replaced by the designator "onset of puberty" which might be a better predictor of growth spurt in other cultures or settings than the particular ages involved in the initial observations. That still would leave questions concerning the factors regulating the onset of puberty. In this example, one can see that as a theory reaches for greater explanatory power, it uses differences to deduce the meaning of the nominal variables from which it has begun. The goal is to relate performance to variables that can, in principle, be manipulated.

Overall, because the kinds of processes or abilities that students of memory wish to make inferences about are hypothetical constructs (such as memory), it is almost a necessity to employ a variety of paradigms and to conduct analyses in a theoretical context. The following sections describe some phenomena of animal memory and then present and evaluate one particular theory of memory. Both the successes and the failures of the theory may serve to illustrate the merits of a greater interplay between theory and data in animal memory research.

PROBLEMS FOR AN ANALYSIS OF ANIMAL MEMORY

We teach a rat a task over a period of days and we note that the rat's performance typically drops off between the end of one session and the beginning of the next session. We also note that when some irrelevant aspect of the learning situation is changed, performance shows an abrupt deterioration. On another occasion reinforcement is withheld (an extinction procedure) until the learned behavior no longer occurs but the next day the animal again persists in the learned behavior. Other experiments show that the rate of forgetting can be altered by either prior learning or experiences intervening between the time of learning and the retention test.

What is the relationship between these observations of spontaneous regression, stimulus change decrement, spontaneous recovery, proactive interference, and retroactive interference? The most salient goal for theories of animal memory is to relate such distinct phenomena within a single conceptual framework. As we shall see, the stimulus fluctuation model suggests a common link between these various phenomena.

THE STIMULUS FLUCTUATION MODEL

The stimulus fluctuation model is but one of a large set of models falling under the rubric of stimulus sampling theory (see Neimark & Estes, 1967); sharing basic assumptions with those other models. Two of the most basic assumptions involve (1) viewing the stimulus situation as comprised of a set of "stimulus elements" that are treated as abstract entities and (2) proposing that on any one experimental trial only a subset of these stimulus elements may be sampled. The stimulus elements are abstract entities and may well be different internal codings or representations arising from an interaction of the organism with the nominal stimulus and the prevailing stimulus context.

The stimulus fluctuation model is distinct in assuming not all of these elements are available to be sampled on every trial. According to the stimulus fluctuation model, stimulus elements can be in either an available or an unavailable state. Owing to inevitable changes in context, elements fluctuate back and forth between these two states as a function of time. Only those stimulus elements that are currently available become associated with an outcome (i.e., become conditioned). After a learning trial the stimulus elements continue to fluctuate so that some conditioned elements become unavailable and some unconditioned elements become available. Finally, after a sufficiently long time the fluctuation process produces the outcome that the proportion of conditioned elements in the available state equals the proportion of conditioned elements in the unavailable state. Performance is determined by the proportion of conditioned elements that are in the available state. This entire process can be schematized as follows, where 0 represents an unconditioned or neutral stimulus element and C represents a conditioned element:

	Before conditioning	Immediately after conditioning	Intermediate state	Asymptotic state
Available	00 0	CC C	C0 C	C 00
Unavailable	000 000	000 000	C00 000	0C0 C00

Referring to the development of the model in Estes (1955a), one can obtain the following equation for the probability of a correct response, $P(c)$, as a function of time since a conditioning trial:

$$P(c) = \frac{c}{c+u} + \left(1 - \frac{c}{c+u}\right)K^t, \qquad (5.1)$$

where t is units of time, c is the proportion of conditioned elements, u is the proportion of unconditioned elements, and K is a parameter representing the absolute rate of stimulus fluctuation. The first term refers to the asymptotic state of fluctuation and the second term refers to the preasymptotic part of the forgetting process, during which a greater relative proportion of conditioned elements are available. These two terms are roughly analogous with long-term memory and short-term memory, respectively.

In his original application of fluctuation theory Estes was able to derive and deduce support for a number of predictions concerning spontaneous recovery and regression as well as trial distribution phenomena. We shall focus now on the predictions of the model more currently associated with the study of memory.

Mechanisms of Forgetting in the Fluctuation Model

Fluctuation. Forgetting in the temporary state embodied in Eq. (5.1) is conceived to be attributable to inevitable fluctuations in the stimulus situation and the organism with time. This assumes, of course, that stimulus conditions are relatively homogeneous during the retention interval and that no special interfering factors have been produced.

When these conditions are met, data consistent with Eq. (5.1) have been obtained. In one experiment (Medin, 1969) monkeys were allowed to view the brief lighting of one cell of a 4 X 4 matrix of cells and, after a delay interval of 0, 1, 2, 5, 10, or 20 sec, were allowed to respond by opening any one cell in the matrix. A reward was always associated with the cell that had just been illuminated. Performance dropped sharply with delay interval but appeared to be approaching an asymptote six to eight times above the chance level.

Performance also depended on which particular cells were illuminated. Corner cells were recalled about three times as well as center cells and edge cells were intermediate in difficulty. However, each of these cell categorizations produced forgetting curves displaying a rapid drop and approach to an above chance asymptote. Medin (1972b) showed that all three forgetting curves could be fit very well by Eq. (5.1) and that the differences between the curves were associated with differences in the long-term memory parameter $c/(c + u)$ and not differences in fluctuation parameter. The pattern of results conforms nicely to expectations based on the fluctuation process in the model.

Context change and retrieval failure. Perhaps one reason the study of forgetting in animals has not been more extensive is that early investigators have observed so little of it. Although contemporary analyses have moderated strong claims concerning the absence of forgetting (e.g., Gleitman, 1971; Spear, 1971), it is the case that animals trained in a situation, given little interfering training during the retention interval, and tested in a situation identical in all respects to the training situation may demonstrate remarkable retention from months (Skinner, 1938) up to years (Liddell, James, & Anderson, 1934).

However, what happens when the training and test situations are not identical? According to the fluctuation model, forgetting may be produced by changes in context because such changes introduce new, unconditioned stimulus elements as well as remove some conditioned stimulus elements.

Contextual variables do seem to alter retention. As early as 1917, Carr observed that performance on a spatial discrimination in a maze was lowered by (1) increases and decreases in illumination, (2) a change in either the position of the maze or of the experimenter in the room, or (3) rotation of the maze. Using a classical conditioning paradigm Girden (1938) found that forepaw flexion (the CR) to a buzzer (CS) depended on the location of the buzzer in the chamber. Spear's (1971, 1973; Spear & Parsons, Chapter 6 in this volume) work on retrieval failure can be interpreted naturally in this framework because contextual elements can be thought of as retrieval cues.

Interference. In typical interference paradigms one trains on a Task A, then trains on Task B, and finally tests for A (assessing retroactive interference) or tests for B (assessing proactive interference). The predictions of the fluctuation model are clearest when Tasks A and B are given in the same situation with the same stimuli, because then one can assume the stimulus element population is largely the same in either task (presently we shall consider the case where they are different). The following diagram conceptualizes the model's predictions for an experiment where extended spaced practice on Task A is followed by massed practice on Task B.

Tests on Task B assess proactive interference. An appropriate control group would not have received training on Task A and the unconditioned stimuli may be viewed as "neutral," that is, being associated with neither Task A nor Task B

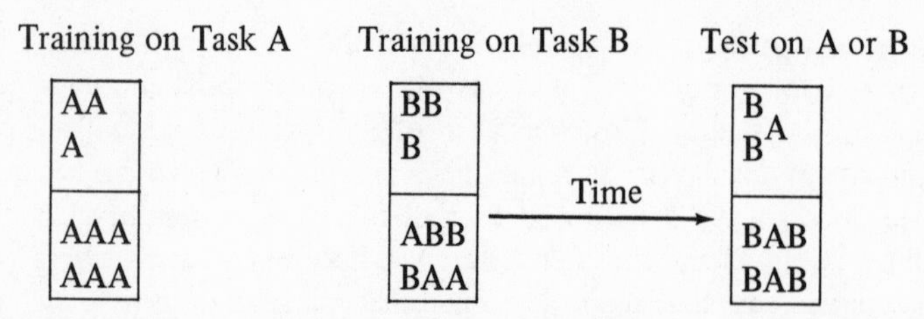

response. If these neutral elements have no influence on whether responses appropriate to Task A or Task B occur, then a number of predictions concerning proactive interference can be derived. The stimulus fluctuation model predicts proactive interference to increase with training on Task A, decrease with training on Task B, and increase with the interval between Task B training and the test. The first two predictions are in general accord with data (see Gleitman, 1971; Spear, 1971, for reviews); there is also evidence that proactive interference increases with time (Frankmann, 1957; Gleitman & Jung, 1963; Grant, 1975b; Koppenaal & Jagoda, 1968; Zentall, 1970; Zentall & Hogan, 1974).

Tests on Task A in the above paradigm provide a measure of retroactive interference. According to stimulus fluctuation theory, retroactive interference is produced by a counterconditioning of elements on Task B. A possible situation for studying retroactive interference may be where Task A is going left in a T maze and Task B is going right in a T maze. As Winograd (1971) and Spear (1971) have indicated, the animal has no way of knowing whether the experimenter wants it to perform appropriate to Task A or to Task B on the test. Unless there are cues distinguishing Task A and Task B, an animal is left to its own devices.

In practice, situations A and B may not be identical. The fluctuation model predicts that as the similarity of the two situations decreases interference should decrease. Wickens, Hall, and Reid (1949) found that switching between hunger and thirst on a T-maze reversal facilitated reversal learning and reduced retroactive interference when animals were switched back to their original habit. Chiszar and Spear (1969) and Zentall (1970) reported that in situations where Task B was the reversal of Task A, retroactive and proactive interference could be dramatically reduced if contextual cues were altered so as to distinguish the two tasks. Even when Task B is not a reversal of Task A, retroactive interference is greater, the greater the similarity of the relevant stimuli in the two tasks or situations (e.g., Zentall, 1973). Finally, Glendenning and Meyer (1971) observed that retroactive interference is controlled more by similarity of motivational states (food vs shock motivated) than by temporal contiguity (see also Meyer, 1972; Howard, Glendenning, & Meyer, 1974).

A serious task for animal memory researchers is to disentangle the various procedures and paradigms where interference does and does not appear. Consider, for example, the delayed-response and delayed matching to sample paradigms, which are often used in studies of primate memory. In the delayed-response paradigm subjects are cued by one means or another as to the location of a food reward and, after a delay, are allowed to respond to any one of a number of locations, including the location where the reward was placed. Usually there are just two locations (the left- and right-hand foodwells of a food tray). Subjects do not respond while the cuing takes place but after the delay can obtain a reward by responding to the previously cued location. The correct or cued location is varied from trial to trial. Yerkes and Yerkes (1928) described the delayed response task as involving primary reinforcement from the previous trial competing with secondary reinforcement from the current trial. From this description one may expect considerable interference, and some evidence of interference has been obtained (e.g., Harlow, Uehling, & Maslow, 1932). Yet, overall, it must be said that proactive events are not a major source of forgetting in spatial delayed response (for a recent summary, see Medin & Davis, 1974).

In contrast, delayed matching to sample procedures are associated with powerful interference effects. In this paradigm subjects respond to a sample stimulus and receive a reward and then, after a delay interval, are given a choice between that stimulus and alternative stimuli, and responses to the sample stimulus are rewarded. Because the correct stimuli change from trial to trial, the incorrect stimulus on a trial may have been the correct stimulus on the just preceding trial. D'Amato (1973) has suggested that this source of interference is so strong that the basis of performance in this task can be thought of as a temporal discrimination where subjects try to remember which object has most recently been rewarded. Even so, a comparable argument can be made for the delayed response paradigm, so it is not obvious what the basis of the differences in susceptibility to interference is.

"Proactive interference" is being used in a descriptive, rather than explanatory, sense. Presumably, processes assumed in the fluctuation model may be used to derive the circumstances governing the presence or absence of interference. The most obvious candidate for this analysis is the similarity of the trial baiting and test conditions in relation to conditions associated with preceding (and potentially interfering) trials and events.

One difference in procedure between delayed response and delayed matching is that the opaque screen of the WGTA normally is lowered during the delay for delayed matching but not for delayed response. Lowering the opaque screen also signals the end of a trial, and the confusion between lowering the screen for a delay and for the end of a trial may be a source of errors. Some support for this interpretation was supplied by a study conducted by Motiff, DeKock, and Davis (1969), which showed that memory performance was improved if a hand-held screen was substituted for lowering of the opaque screen during the delay interval.

Even when the current event and the interfering event are not distinguished by distinctive stimuli, such as the lowering of the opaque screen of the WGTA, it is hard to demonstrate proactive interference in delayed-response paradigms. Frank

Ruggiero, Roger Davis, and I recently ran an experiment using a 4 X 4 matrix of cells where the lighting conditions involved two distinct colors. A given trial was one of three types:

1. C_1–delay interval–test (subject must remember the position of C_1, which is randomly assigned on each trial to one of the 16 positions)
2. C_2–delay interval–test (remember C_2)
3. C_1–interstimulus interval–C_2–delay interval–test (remember C_2)

The third type of trial provides a test of proaction. Subjects must attend to the C_1 light (because the trial may be of the first type) but, when the C_2 light appears, must ignore the position of the C_1 light and remember the position of the C_2 light. By comparing Trial Types 2 and 3 one can assess interference effects. A possible confounding is that subjects may not even look at the second light for Trial Type 3, and one may not be able to separate attention from interference effects. However, one can adjust for differing attending probabilities because a 0-sec delay interval is included. Monkeys were tested for delays of 0, 1, 3, or 12 sec and the interstimulus interval was also either 0, 1, 3, or 12 sec. Subjects were tested 48 trials each day for 30 days.

The results revealed a reliable but extremely small interference effect, even after the scores were adjusted for differing probabilities of attending to the second light. The relative lack of interference on the second light by the first light cannot be explained by suggesting that memory for the first light is absent at the time that memory for the second light is tested. Intrusion errors (responses to the location of the first light) increased with delay interval, which was consistent with the idea that when information concerning the second light was forgotten, subjects responded to their information concerning the location of the first light. However, the observed interference effect was too small even to account for the number of these specific intrusion errors.

Although nominally, subjects were rewarded for choosing the most recently illuminated cell location, the color of the cell was a redundant relevant cue and the availability of the color cue might have been responsible for the relative lack of interference. In contrast, if the lights were not of distinct colors and subjects were required to respond to the location of the most recently illuminated cell, the fluctuation model would predict that much stronger interference would be observed.

Still another difference between delayed response and delayed matching is that in delayed matching a response is made and often a predelay reward is given but no response is made and no predelay reward is given for delayed response. One might expect such a difference to favor delayed matching, but Cowles (1941) found a predelay reward produced a dramatic drop in delayed response performance in rats. I have attempted several times to use a delayed matching to position procedure involving reward on the sample trial (essentially Cowles' paradigm), varying delays, intertrial intervals, and even the number of sample trials without producing performance much above chance with pigtailed monkeys.[1]

Repetition Effects

If we repeat conditioning trials, some predictions of the fluctuation model in relation to the spacing of trials become obvious. A trial immediately following a conditioning trial would be inefficient because the only stimulus elements available to be conditioned would have already been conditioned. A trial is most effective if it is spaced because some unconditioned, unavailable stimulus elements have a chance to become available.

The experimental literature on spacing effects in animal memory is sparse but of interest. The most systematic report to date on spacing and repetition effects involved the use of pigeons in a delayed matching to sample task (Roberts, 1972b). In this paradigm birds are presented with a single (sample) stimulus and then, later on, must select that stimulus from a set of alternative stimuli. A series of careful studies failed to show any evidence of an advantage for spaced repetitions of the sample stimulus and uniformly supported a simple trace strength decay model. This model assumes that memory can be represented by a single memory trace, which may vary in strength. This strength is increased by repetitions and, during retention intervals, owing to either time decay or interference, decreases at some constant rate toward zero strength. If the retention interval between the second presentation of an event and a test is held constant, this theory predicts performance to improve the closer the first presentation is to the second, because less forgetting of the first presentation has occurred. Roberts observed a massing advantage and was led to speculate that very different mechanisms of STM are at work in man and bird.

Seemingly, these results are in direct contradiction to the prediction of the fluctuation model. However, in Roberts' studies a random one of a set of four alternative stimuli was presented on the "sample" trial and, for tests, a pigeon was given a choice between the sample stimulus and one of the other stimuli. On the next trial, some one of the other stimuli might be the sample, and over a series of trials all of the stimuli were associated with reward. The pigeon must therefore remember which of the stimuli has most recently been paired with rewards, because the incorrect alternative on a particular trial may well have been correct on the just preceding trial. In this sense, the procedure resembles the interference paradigms discussed earlier.[2] Compared to a situation where stimulus elements are either conditioned to the correct response or else neutral, spacing advantages become much smaller or absent where elements are either

[1] William Roberts has suggested to me that predelay reward could be having the same effect as lowering an opaque screen, that is, suggesting termination of a trial.

[2] This speculation must be tempered by the fact that some direct tests for proactive interference have failed to observe it (see Roberts and Grant, Chapter 3).

conditioned to the correct or to incompatible responses (as in the interference paradigm).

To examine the relationship between interference and the spacing effect, Medin (1974) tested monkeys on a delayed matching to sample paradigm where the stimulus pool varied in size from day to day. On some days stimuli never appeared on more than a single trial and subjects just had to remember which of the objects had been presented and rewarded. On other days only two objects were used, a procedure which should have tended to maximize interference because the correct object on one trial might be the incorrect object on the next trial.

The results conformed to the expected pattern. Whether or not a spacing advantage was observed depended on the stimulus condition: when only two different objects were used on a day, proactive interference was prominent and no difference between massed and spaced repetitions was observed (at the long delays there was even no repetition effect); when the stimuli for each trial were new, there was little or no interference, spaced repetitions led to better performance than massed repetitions of the sample, and this difference increased with delay interval.

Generally this experiment supports the fluctuation model. Actually, modifications in the model seem to be required when one attempts to fit the observed data more precisely and in order to predict the massing advantages that Roberts observed. One possibility is that not all of the available stimulus elements get conditioned as the result of some trial event. One can greatly improve the fit of the model to data without changing any of its qualitative properties by assuming that some proportion, c, of the available stimulus elements become conditioned on a trial.

There really is not sufficient evidence available with animals to get a clear picture of the effects of spaced repetitions. In the only other studies I am aware of, spacing advantages have been observed in both rats (Roberts, 1974) and chimpanzees (Robbins & Bush, 1973).

Summary and Assessment of the Original Fluctuation Model

The stimulus fluctuation model is simple in conception yet generates a considerable variety of predictions that are roughly in accord with the data. To be sure, the model has not been rigorously tested because few sets of systematic data have been gathered. Even so, the fluctuation model has displayed unusual longevity, especially today when new models seem scarcely to outlive their publication lag. However, a number of developments in the last 20 years on a broader plane than just a discrepant result here and there, all told, suggest to me that some basic modifications in the fluctuation model may be in order. In the next few paragraphs I explore a few of these developments and then offer a modified version of stimulus fluctuation theory.

Direct associations. One of the basic tenets of the fluctuation model is that performance is determined by independent, simple associations between stimulus elements and responses (in an S–R framework) or between stimulus elements and events (in an S–S framework). However, there are several kinds of experiments that pose problems for this view. Consider two such experiments.

Riopelle and Copelan (1954) trained monkeys on a series of discrimination shifts where the correct and incorrect stimuli were reversed from time to time. In particular, whenever the color of the food tray on which the stimulus objects were presented was switched, the reward conditions were switched. Monkeys were able to use the color information to perform without error on these discrimination reversals.

Another series of experiments conducted by Wyrwicka (1956) was concerned with the relations between responses trained in different situations. Situation A was a conditioning chamber with a dog being placed on the stand and fed from a feeder; Situation B was a second room with the dog standing on the floor and having bread pieces thrown to him by the experimenter. The dog was trained to make Movement A to Stimulus A in Situation A and to make Movement B to Stimulus B in Situation B. Then the dog was tested with Stimulus A in Situation B and with Stimulus B in Situation A. On these tests the dogs made Movement B to Stimulus A in Situation B and Movement A to Stimulus B in Situation A. In other words, responses were controlled more by the situations than by the stimuli. Adding a nonrewarded Stimulus A′ to Situation A to force differentiation between Stimulus A and Stimulus A′ did not change the pattern of results. Dogs still made Movement B to Stimulus A in Situation B and made no response to Stimulus A′ in Situation B.

How can a simple associative theory handle either set of results? One might resort to redefining the stimulus; in the Riopelle and Copelan study, for example, one could propose that the discriminative cues and the conditional cues (food tray color) came to act as a single compound stimulus or that the set of stimulus elements consisted of compound stimuli which fluctuated as in the original interpretation of the model. Such a mechanism can also be used to explain configural conditioning experiments in which one finds subjects can learn to respond to Components C and D and avoid responding to a CD compound (e.g., Woodbury, 1943). However, one must wonder how gracefully the assumption of component or stimulus element independence can withstand such compromise. Shortly I shall present a modification of fluctuation theory that gives us both the assumption of component independence and the idea that associations are formed directly to cues.

Cue and context. Although it is natural to discuss context change as a source of forgetting in the fluctuation model, fluctuation theory does not distinguish between cue and context as sharply as one may like. The desire to make a sharper discrimination is probably, more than anything else, a personal preference, but it is based at least partly on the goal of relating memory research with theoretical developments in the area of discrimination learning (cf. Medin, 1975).

Memory improvement. One of the most salient findings concerning animal memory is that performance almost always gets better (sometimes dramatically better) with practice (D'Amato, 1973; Medin & Davis, 1974). Even though the fluctuation model may be a useful beginning point in analyzing these improvements, it clearly is not directly addressed to them.

None of the above difficulties with fluctuation theory would likely overwhelm ardent supporters of the theory. However, I shall take these problems as strong hints that exploring some modifications of fluctuation theory may be fruitful. The ideas to be sketched in the following paragraphs borrow heavily from Estes (1973) and Medin (1975).

A MODIFIED FLUCTUATION THEORY

We begin by first drawing the distinction between cue and context and then, to a degree, blurring it. Roughly, cues are those aspects of a situation that the experimenter varies and to which it is assumed subjects respond, whereas "context" refers to those relatively invariant aspects of a situation (from the experimenter's point of view) in which a response occurs. However, a particular stimulus may function both as a cue and as context for other cues.

To illustrate this distinction, imagine that a blue circle is responded to in some experimental context and that some event (say, a reward) occurs following this response. The main proposition is that information concerning both cues and the context in which they occur is stored together in memory. Furthermore, I assume that both the cue and its associated context must be activated simultaneously in order that information about an event be retrieved. Either changes in cues or changes in the context can impair the accessibility of (reinforcement) information associated with the situation. These ideas may be pictured as follows:

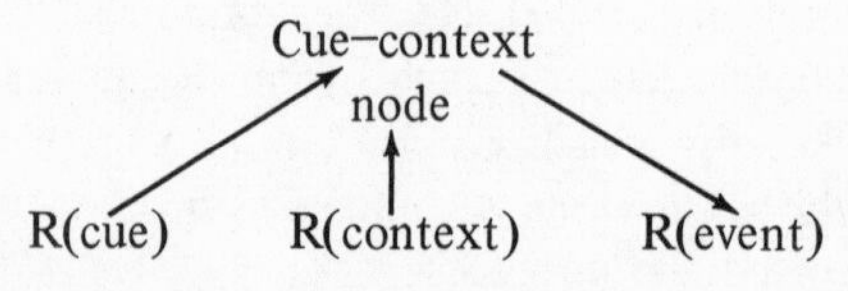

R(cue), R(context), and R(event) refer to the memory representation of the cue, context, and event, respectively. The cue–context node corresponds to what Estes calls a "control element," and I use it to denote the assumptions (1) that neither cue nor context is directly associated with an event or outcome and (2) that inputs from both cue and context are needed to activate the node and provide access to the representation of the event. The latter assumption implies that the effects of cue changes and context changes combine in an interactive manner.

Now consider again Wyrwicka's experimental results. Predictions from the modified fluctuation theory depend on the similarity of Stimuli A and B and the similarity of Situations A and B. If the two stimuli are more similar than the two situations, then the cue–context node that is activated depends mainly on the situation in which the dog is placed. There should be no unusual difficulty for the special case where Stimulus A and Stimulus B are identical, because activation of the cue–context node depends both on cues and on context. The Riopelle and Copelan experiment can be viewed in this manner if we represent the food tray color as a contextual cue.

The link between the above ideas and the fluctuation model is straightforward. We assume the elements that are fluctuating in the modified fluctuation model are the contextual elements. On each trial or presentation in a memory paradigm, each of the available contextual elements has an opportunity to form a cue–context node with each of the cues in the situation. A given cue may be associated with a number of different contextual elements and a given contextual element may be associated with a number of distinct cues. Each cue–context node is unique, however, and at a given moment, only one event can be associated with the node.

The basic mechanism of contextual fluctuation preserves the generally successful predictions of the original model concerning the form of the retention curve, sources of forgetting, and spacing of repetitions. Assuming that a cue–context node is formed with some probability, which may be less than unity, can account for the occasional observation that massed repetitions yield better performance than spaced repetitions at short retention intervals. (Recall that a parallel modification in the original fluctuation model seemed to be needed.)

In both the original and modified form of fluctuation theory the assumption that all elements fluctuate at the same rate is a convenient abstraction rather than a strong conviction. In actuality, certain classes of contextual cues may fluctuate at different rates. For example, Krane and Wagner (1975) report findings consistent with the ideas that after a delay interval gustatory stimuli are more available than visual stimuli. Some of the unusual findings in the taste-aversion learning literature may be consistent with this possibility (see Bolles, Chapter 2).

Little has been said so far about stimuli, except that changes in a cue (such as for generalization tests) also decrease the probability that a cue–context node becomes activated. Following my earlier theoretical work (Medin, 1975) I propose (1) that cue differences along each stimulus dimension can be represented by a similarity parameter ranging in value between 0 and 1, with 1 representing complete identity and 0 representing no similarity; and (2) that stimulus change decrements along the various cue dimensions combine in a multiplicative manner, yielding a single similarity measure governing transfer or retrieval. The latter assumption may be contrasted with the more usual assumption that transfer is governed by the average similarity of two situations.

The multiplicative rule is consistent with the idea that a stimulus value may act both as a cue and as a context for other cues. These assumptions have led to straightforward predictions concerning the relative rates of learning various types of discrimination problems, and they can also be tested in memory paradigms varying the similarity of the correct and the incorrect (distractor) stimuli.

The next few paragraphs describe a preliminary experiment directed toward the assumption that cue and context changes alter performance in an interactive (multiplicative) manner. More detail than usual is given because this experiment is not reported elsewhere.

Context Change and Proactive Effects

Proactive interference effects are very prominent in delayed matching tasks where a small set of stimuli is used. In this study we attempted to examine these effects more systematically. On each sample trial monkeys saw either a triangle or a circle and then, after a delay interval, were given a choice between a triangle and a circle. Across trials either object was used as the sample for a random half of the trials and it was possible that the incorrect form on a particular trial was the correct form on the immediately preceding trial. Three types of changes between consecutive trials were investigated. The correct object on a given trial was either the same form or a different form from that of the preceding trial, it occupied either the same or different position between trials, and the stimuli were either the same color or a different color between consecutive trials. These conditions are illustration in Fig. 5.1. The main interest is in the relationship between the effects of these changes; the modified fluctuation model predicts these factors to interact.

Method

Subjects. The subjects were five jungle born pigtailed monkeys (*Macaca nemestrina*), 4–5 years old. All animals had received approximately 2 years of training on various object discriminations, including delayed matching.

Apparatus. A Wisconsin General Test Apparatus was used for testing. The three foodwells of the grey food tray were spaced 15-cm apart center to center. The center foodwell was always used for sample presentations, whereas the two side foodwells were used for tests. The stimuli consisted of two equilateral triangles and two circles constructed from 1.2-cm plywood. One circle and one triangle were painted white and the other two stimuli were painted black.

Procedure. The monkeys were tested 5 days a week for a total of 30 days on delayed matching to sample problems. On each trial the sequence consisted of

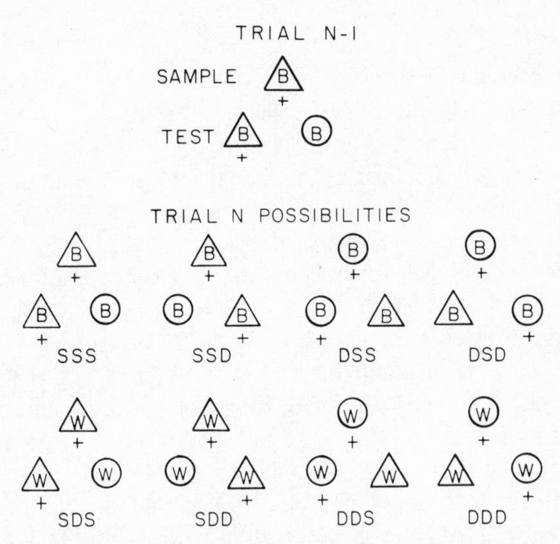

FIG. 5.1 The eight possible relationships between the correct stimulus on the preceding trial and the correct stimulus for the current trial. Pairs of such stimuli may have either the same (S) or different (D) form, the same or different color, or occupy the same or different position.

(1) the presentation of a sample object over the center foodwell, to which the subject responded for a raisin reward; (2) a delay interval of 6, 12, or 24 sec; followed by (3) a two-choice test trial, with two objects appearing over the side foodwells and reward always available under the object that had been used as the sample. The delay intervals and the 12-sec intertrial interval were measured from the lowering of the opaque screen of the WGTA after a response to the raising of the screen for the next response.

The two choice objects always were identical in color and also matched the color of the sample stimulus within a trial. The combinations of the various relationships between two consecutive trials produced eight experimental conditions; on two consecutive trials the objects were either the same color or a different color, either the same form or different forms were correct, and the test stimuli either occupied the same or different positions between trials (see Figure 5.1). Each of the eight conditions produced by the possible combinations of these binary variables appeared three times each day, once at each of the three delay intervals. Because one of the main variables of interest was performance as a function of preceding trial conditions, a "dummy trial" consisting of a random one of the 12 stimulus-delay conditions was given on the initial trial of each day. In all, then, 25 trials were given each day for each animal. Across blocks of 3 days the delay interval on the trial preceding the various experimental conditions was also balanced. Order of conditions within a day was also randomized but no stimulus object or position was correct for more than four consecutive trials. A day's test session lasted approximately 35 min.

Results and Discussion.

Performance decreased from 63% correct to 54% correct from the short to the long delay interval, but the largest effects were attributable to between-trial relationships. Correct responses ranged from 76% correct in the condition when the same object was correct, occupying the same position, and of the same color on consecutive trials, down to 32% correct when a different object was correct, occupying a different position, and of the same color on two consecutive trials. Percents correct for the eight main conditions as a function of delay interval are shown in Table 5.1.

Statistical tests showed that the main effects of delay interval ($F = 8.90$, $df = 2/8$, $p < .01$), same vs different form ($F = 18.91$, $df = 1/4$, $p < .05$), same vs different position ($F = 26.67$, $df = 1/4$, $p < .01$), and the interactions of color and form ($F = 34.57$, $df = 1/4$, $p < .01$) and position and form ($F = 8.56$, $df = 1/4$, $p < .05$) were all statistically reliable.

The interactions of changes in color, form, and position are of greatest theoretical interest. First, a word of caution concerning the interpretation of interactions is in order. Whether or not an interaction is obtained depends strongly on the dependent variable. A transformation of the data (e.g., from proportion correct to log of the proportion correction) may either produce or remove an interaction. Therefore, unless the dependent variable is dictated directly by the theory under consideration, no claim can be made concerning the presence or absence of an interaction, unless the interaction actually produces a crossover (in which case transformations do not remove the interaction). The modified fluctuation theory is developed in terms of probabilities and therefore, for purposes of assessing interactions, proportion correct is the proper dependent variable.

A summary of the data relevant to the predicted interactions is shown in Fig. 5.2. The left panel illustrates the strong interaction of color and form; changes in color impair performance if the same form is correct on two consecutive trials but greatly facilitate performance if different objects are correct on consecutive trials. A change in color does not interact with position, as the middle panel shows. Finally, position and form do interact (right panel); the effect of same vs different form is larger when a shift in the position of the correct object occurs than when a shift does not.

Although the original fluctuation model does not predict any interactions of the variable shown in Fig. 5.2, the modified fluctuation model predicts interactions in all three panels of the figure. Neither result obtained. In the next few paragraphs a straightforward interpretation of these data is offered that is consistent with the modified fluctuation theory and that also suggests that two relatively independent sources of interference operate in the task.

Interference from prior trials may arise from the association of other stimuli with reward and from the specific (positional) response that may be associated with a reward. Compared to the former source of interference, interference from

TABLE 5.1
Percentage of Correct Responses as a Function of Delay Interval and the Relationship between the Current and the Just Preceding Trial

Consecutive trial relationships			Delay		
Object	Color	Position	6 sec	12 sec	24 sec
Same	Same	Same	.81	.82	.67
Same	Same	Different	.69	.59	.69
Different	Same	Same	.60	.50	.44
Different	Same	Different	.35	.30	.31
Same	Different	Same	.76	.63	.63
Same	Different	Different	.58	.62	.53
Different	Different	Same	.73	.62	.61
Different	Different	Different	.49	.50	.45

the prior response may depend much less strongly on the presence of particular external contextual stimuli. Indeed, the prior response may be controlled more by proprioceptive and other such internal stimuli (see Mackintosh, 1974, for related discussion). In the present experiment we propose that position interference reflects a tendency to respond to the rewarded position independent of the contextual cues of color and form, whereas form interference depends strongly on the experimental context. The form by position interaction arises because a change in position alters the effect of same vs different form and not the reverse. That is, it is plausible that effects attributable to same vs different position do not depend on form or on color (middle panel). In contrast, the effects of same vs different form depend both on color changes (left panel) and position changes (right panel). Additional support for the idea that there are two different kinds

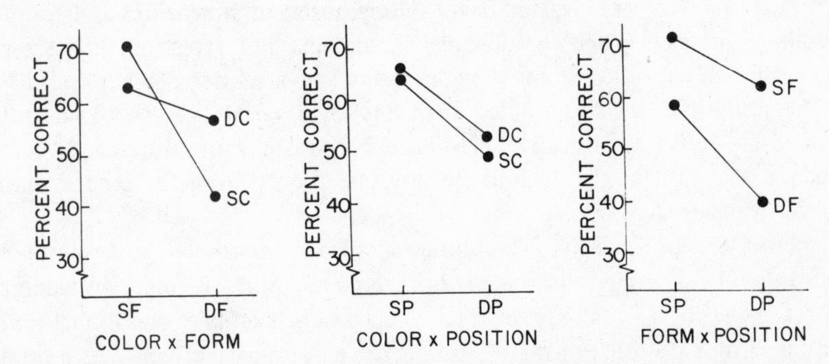

FIG. 5.2 Percent correct responses in relation to the similarity of the correct stimulus on preceding trial to the correct stimulus for the current trial. The two-way interactions of color and form, color and position, and form and position are shown in the left, middle, and right panels, respectively.

of interference in this paradigm comes from the finding that the same vs different position effect decreases with delay interval, whereas the effects of same vs different form do not ($p < .05$ by a sign test).

A further test of the above interpretations of the obtained pattern of interaction may be provided by an experiment where color is the relevant matching cue, whereas form remains constant within but varies between trials. In such an experiment one would expect (1) color and form to interact as before; (2) that now color and position would interact, because changes in position should alter the effect of same vs different color; and (3) that form and position should no longer interact because form would no longer be relevant and position interference would not depend on whether same or different forms were used on consecutive trials. In any event, the present experiment shows that changes in color, form, and position do not have simple additive effects.

Further Assumptions and Predictions

The modified fluctuation theory may also be of use in analyzing improvements in memory performance with practice. In the analysis of discrimination learning phenomena, it proved necessary to assume also that selective attention was a major component in learning (see also Sutherland & Mackintosh, 1971, for an extensive review). Perhaps animals can learn to attend selectively to those cues which are likely to support retention. If so, the improvements in memory with practice may reflect the use of an attentional process. D'Amato (1973) has suggested that learning to attend to temporal cues underlies the improvement his monkeys show on delayed matching trials.

How does this selective attention operate? We shall assume that, given two cues along a dimension, generalization (effective similarity) along the dimension is less when a subject attends to that dimension than when the subject fails to attend to that dimension. Although this assumption is weaker than the more usual idea that no generalization occurs along observed dimensions and complete generalization takes place for dimensions that are not attended, it has served successfully in accounting for a wide variety of attentional phenomena. The weak assumption can be coupled with any of a variety of specific attention theories to derive quantitative predictions; but the current research picture probably does not justify singling out any one theory from the general class of selective learning theories.

If memory improvements are a function of an attentional process, we may make theoretical progress in our investigations of animal memory by analyzing transfer between various experimental settings. For example, one robust finding in the discrimination learning literature is the easy-to-hard effect, where training subjects on an easy discrimination (say, black–white) before a hard discrimination (say, light grey–dark grey) is often more effective than the same amount of training on the hard discrimination (for a review, see Sutherland & Mackintosh,

1971). In an analogous memory experiment Mishkin and Delacour (1975) report that training monkeys on delayed matching problems where familiarity and novelty are salient and relevant yields rapid acquisition of matching behavior. Furthermore, these training effects transfer to and eliminate much of the proactive interference associated with delayed matching problems using just a few alternative stimuli, where relative familiarity is much less salient.

Summary of the Modified Fluctuation Model

The modified fluctuation model preserves most of the predictions of the original fluctuation model yet overcomes many of the difficulties arising from the original assumption that responses become directly conditioned to stimulus elements. Data on the interaction of cue and context changes are consistent with the posited cue–context node. The model seems applicable to studies of retrieval and retrieval failure; potentially it may also serve as a tool in analyzing improvements in memory performance. We need both more theoretical and more empirical work on many such issues.

CONCLUSIONS

Having spent considerable time discussing particular memory models, I return now to the issue of animal models, especially in relation to the study of cognitive processes. Early in my training I supposed that the relationship between human and animal research was roughly as follows: (1) animals were more simple, and therefore the basic principles of behavior should be easier to deduce in animals; (2) animal and human research might just as well proceed independently, because we could not compare the populations intelligently without already having in hand the very knowledge we were seeking; and (3) that someday, sometime in the future, we who worked with animals would come up with the answers and then reveal the truth to those who had been working exclusively with human subjects.

All three of the above tenets foster a misleading view of the relationship between human and animal research, in my opinion. Animal cognitive functioning may in fact be more simple, but it is at least as relevant that animal cognitive functioning may be different—that is, animal research may be a source of variability that can serve to develop, differentiate, and test theories. Therefore, if monkeys, which are unable to produce human speech sounds, nonetheless show categorical perception of these same speech sounds, direct implications for motor theories of speech perception will be obvious (Morse & Snowdon, 1975). Likewise, who would not be intrigued and ready with suggestions for future research if short-term memory in pigeons could be shown to be fundamentally different from short-term memory in rats, monkeys, and man?

Relationships between human and animal research can be fruitful when there is a direct ongoing interaction. For example, the fluctuation model's prediction that proactive interference diminishes the spacing effect has implications for human research that I am currently examining with college students. Jahnke (1974) has already reported that the magnitude of the Ranschburg effect, having to do with poor recall of repeated items, is related to the size of the stimulus pool employed and the resulting proactive interference.

To amplify this kind of interaction, we need considerable work on animal memory models. As a partial step in this direction I have analyzed the original stimulus fluctuation model and offered a modified fluctuation theory. The present framework contains many of the concepts developed by Estes in his hierarchical associative model for memory, which has been applied to topics ranging from memory for order of spoken letters to basic classical conditioning phenomena (Estes, 1972, 1973). It is also of more than passing interest that Bower's (1972) adaptation of Martin's theory of encoding variability embodies the fluctuation process of the original Estes model. Collectively, these observations may forbode greater interplay between human and animal work on a theoretical and empirical level.

Finally, we turn to the relationship between animal memory models and the interest in animal models as a means of revealing the relationship between brain structure and behavior. If a particular lesion results in impairment in delayed-response performance, what can be said? One guess is provided by the following quotation: "Psychologists have often been at a loss as to how to classify the delayed response. It has been variously considered as a learning problem, attention problem, memory problem, and thinking problem" (Harlow, Harlow, Ruepling, & Mason, 1960, p. 120). Because much physiological work involves relating brain structures to processes that are hypothetical constructs, such as memory, these efforts have an inherent stake in the development of better theory.

The future holds the answer to whether or not behavioral research with animals can facilitate either physiologically oriented endeavors or research with human subjects. My opinion is that even in the study of cognitive processes we can fruitfully imitate other natural science disciplines in exploiting animal models to further our understanding of humankind.

ACKNOWLEDGMENTS

This research was supported by United States Public Health Service Grants MH 25134 and MH 16100. This paper was written while the author was a visiting assistant professor at the University of Michigan. W. K. Estes, Donald Robbins, and William Whitten read earlier drafts of this paper and provided helpful suggestions.

6

Analysis of a Reactivation Treatment: Ontogenetic Determinants of Alleviated Forgetting

Norman E. Spear
Patricia J. Parsons

State University of New York, Binghamton

INTRODUCTION

We readily can agree that forgetting, operationally defined as a decrement in the manifestation of learning, is a common feature of behavior and that if certain "reminder" events occur, forgetting may be overcome. The purpose of this chapter is to examine a relatively restricted but powerful case in which forgetting regularly is alleviated by presenting a "reminder-like" event. Because of our particular point of view, we prefer to describe such a reminder-like event as a "reactivation treatment." After we discuss the generality of reactivation treatments and how their effects may be explained in a broad sense, we shall trace the analysis of one reactivation treatment, with particular attention to how its effects may differ for animals in different ontogenetic stages.

THE SCOPE OF REACTIVATION TREATMENTS

A variety of sources of forgetting may be identified operationally. For example, forgetting may be induced because of prior or subsequent learning that apparently is incompatible with the memory being tested.

In such operational terms, "interference" refers to the detrimental effect on retention resulting from the acquisition of memories that in some sense conflict with the memory being tested; conflicting memories may have been acquired proactively (prior) or retroactively (subsequent) to acquisition of the critical memory. Perhaps the most familiar source of forgetting operationally is a retention interval interpolated between acquisition and testing, the effect of which may be attributed to extraexperimental interference, a decay process, change in contextual stimuli, and so forth. Somewhat less common is an insult to the central nervous system by electrical or chemical events that disrupt normal neurophysiological functioning. Of key interest here is "retrograde amnesia," in which a physiological insult following perception of events to be stored as memories results in forgetting of those events. Still another source of forgetting may be aging and its correlates, especially including those neurophysiological changes in an organism that occur in regular sequence in the course of maturation and so may develop between acquisition of a memory and the test for retention.

Whether or not one took the view that a common process underlay these descriptively distinct sources of forgetting, it would be of some interest if the effects of each source were similary reduced by certain reactivation treatments. This possibility has tended to be confirmed (Spear, 1973). Furthermore, analogous instances of this relationship have appeared both in studies using human subjects tested on primarily verbal memories and in studies with animal subjects tested on instrumental or classical (Pavlovian) conditioning. We now may present briefly a few specific examples of such studies that fall roughly into the categories of clinical evidence, human verbal memory processing, and the psychobiology of memory processing in animals.

Clinical Evidence

Alleviation of forgetting caused by a variety of sources—from explicit brain injury to implicit psychoanalytic repression—is a common aspect of clinical treatment. Barbizet (1970) relates three means typically applied so that impaired recall may be improved following violent cranial trauma. One treatment is the simple application of general recall exercises—merely asking the patient to remember certain episodes. Such exercises, Barbizet suggests, may have the effect of permitting the patient to realize his memory disorder, and may act through some unknown mechanism to "induce a better reconstruction of the past" (p. 127). A second source of aid is information about prior events provided by friends and relatives of the patient. The third is facilitation through reexposure to events that have formed portions of memories acquired in the past.

> Mr. D, 5 months after an accident that had produced one month of coma, had lost all memory of his personal life over the preceding 20 years. Two months later, while on leave from the hospital, he went back with a cousin to his old home. Here he found himself remembering the address, the layout of the place, the name of his neighbors, etc., and these memories persisted. Mr. F., four months after his cranial trauma, had entirely forgotten what he had been doing during the two months preceding the

accident. Six months later he was able to relate that he had in this period bought a car, rented an apartment, but he had to visit the garage and the apartment before he was able completely to recall his memories [Barbizet, 1970, p. 128].

Williams (1969) states simply that "All observers have noticed that RA (retrograde amnesia) shrinks in the early days of recovery and that shrinkage tends to be helped by cues, prompts, and revisits to familiar places [p. 77]." Toward developing a theory of memory processing on the basis of such clinical cases, Barbizet (1970) suggests that permanent loss of the physiological substrate of a memory may be prevented "by an unconscious reactivation each time the the subject experiences an element of the repressed conflictive situation [p. 146]."

Therapists involved in the treatment of aphasia (generally, a disturbance in verbal communication) are well aware of the importance of events that may serve as retrieval cues to aid patients in the recall of particular verbal items. Hecaen (1969) has noted that for patients suffering the form of aphasia characterized by simple failure to provide the appropriate words during conversation (termed "amnesic aphasia" by Hecaen), the deficit is most obvious in circumstances under which the words required are unconstrained by syntactical or semantic context. Eisenson (1973) points out the accepted premise of aphasiologists that the nature of their clinical therapy is not to teach the patient anything new, "but to help him retrieve what he knows more readily than if he were left to his own devices [p. 134]." Later, Eisenson notes that "a patient who cannot readily evoke the name of a number can be trained to count serially until that number comes up in a sequence, and then to stop at it" (p. 158). In this way, presumably because they are provided a sequential linguistic context that aids retrieval of the particular number, aphasic patients learn to say their telephone numbers, addresses, and other critical information.

Demonstrations of what appears to be control over retrieval of memories under paraexperimental circumstances have been described by Tompkins (1970) and by Blum, Graef, Hauenstein, and Passini (1971). Tompkins elaborates on the notion that if therapists are interested in facilitating the recall of childhood memories in their patients, they should expose their patients to events common to childhood. Tompkins reported an experiment in which he instructed subjects to write their signatures very slowly, taking, in childlike fashion, approximately 3 sec to write a single letter, and a second experiment in which his subjects were required individually to stand in front of a group and shout "No, I won't!" Tompkins presents no detailed data but reports that among the subjects in both experiments who complied with the instructions, 95% gave evidence of retrieving early memories. The subjects who slowly wrote their signatures tended to adopt childlike handwriting styles, held the pencil tightly "as they did earlier," and reportedly retrieved ". . . early memories of the first-grade schoolroom and the first-grade teacher [p. 88]." Subjects in the other experiment were observed to protrude their lower lips immediately after shouting, similar to a pouting child, and they reported experiencing childlike emotions and recollecting long-forgotten incidents.

Blum *et al.* (1971) report what appears to extraordinarily good retention of an imaginary episode that had been presented to a subject under the influence of an induced "distinctive mental state." The good recall apparently was contingent on the recurrence of the state, which was created by suggesting to the hypnotized subject that he was floating in a "beautiful blue mist" 2 ft above the couch on which he was lying and could "feel a light breeze blowing." The subject then was presented a bizarre, imaginary episode consisting of President Johnson (this was during 1967) leading a peace march in Viet Nam but wearing the face of Sammy Davis, Jr. A retention test was given quite unexpectedly after a 143-day interval. Blum *et al.* (1971) reported the subject describing the entire episode with accruacy and a somewhat incredulous tone in his voice on reestablishment of the "distinctive mental state" under which the episode had been learned.

Obviously, these sorts of reports must be accepted as being on little more than anecdotal level; the interpretative problems are innumerable. For example, in a study related to the one reported by Tompkins, Orne (1951) hypnotically regressed an individual to age 6 years and found that the individual did print in the style of a child of that age but also printed, with perfect spelling, the phrase, "I am conducting an experiment which will assess my psychological capacities [p. 219]." What is clear, however, is the common clinical assumption that previously acquired memories may be retrieved on the occurrence of events that are perhaps best described as contextual in nature rather than as specific discriminative or conditioned stimuli.

Human Verbal Memory Processing

There are three general areas of treatment that tend to enhance retrieval of verbal memories: these treatments involve the influence of semantic context on free recall, the influence of nonsemantic environmental context on retention, and the alleviation of retroactive or proactive interference by semantic cues. We shall describe briefly these effects.

When a set of 20 or so words is presented and the subject is asked to recall as many as possible in any order, the total number of words recalled may be increased by providing the subject with certain semantic cues as part of the context of recall. When subsets of words can be grouped easily into categories (for example, COW, HORSE, and DOG may be categorized as animals; LETTUCE, CARROTS, and BEANS as vegetables), the benefit to recall will be especially effective if the names of the categories are provided the subject. The actual number of words recalled may be increased as much as 50–100% with this operation, with greater benefit when the names of the categories had been presented explicitly with the other words in the list. When words in the list do not fall readily into natural categories, cue words from outside the list may facilitate recall if presented at that time, especially if they correspond to some scheme by which the subjects organize the words for recall (see Wood, 1972; Tulving & Thomson, 1973; Watkins & Tulving, 1975). However, if some of the words that actually appeared in a list are presented at recall of that list, performance usually is not facilitated and in fact may be impaired, perhaps as a consequence of "output interference" (see Roediger, 1974; Rundus, 1973).

Nonsemantic features of the testing environment also may determine retention as a function of how faithfully these features correspond to those during original learning. McGeoch (1942) cited this effect as one of the two most important factors determining retention. He based this judgment partially on rather incomplete evidence that indicated poorer retention of classroom learning if the testing room and proctor were different from those during learning, and partially on data that indicated that persons who learn a list of words in seclusion show better retention if tested in seclusion than if tested with an audience. Greenspoon and Ranyard (1957) introduced changes in the specific room and physical features associated with the learning and testing of two consecutive lists of verbal units; in one room, bright and cluttered, the subject stood throughout the experiment, and in the other room, dark and relatively empty, the subject sat in a soft chair. Their design permits one to subtract statistically the effects of different intervening treatments in order to compare retention of the first list when the training environment is replicated at testing relative to retention when the environment is different at testing (see Keppel, 1972). Through this procedure one arrives at the estimate that retention is approximately twice as good under the conditions where the physical circumstances of original learning are replicated during the retention test. Similarly, Rand and Wapner (1967) had subjects learn a list of words while either lying on their backs or standing up, and then tested retention when they were either in the same position or in the alternative position. Retention tended to be better when physical posture was the same during testing and learning.

Finally, forgetting caused by experimentally induced retroactive interference has been alleviated markedly by presenting, at recall, semantic cues, such as names of the categories or specific items associated with the original learning (e.g., Tulving & Psotka, 1971). Comparable alleviation of proactive interference also has been reported as a consequence of presenting certain semantic cues during testing (Gardiner, Craik, & Birtwistle, 1972).

Alleviation of Forgetting in Psychobiological Research

Forgetting in animals may be attributed to a wide variety of sources: long retention intervals; retroactive and proactive interference; warmup decrement; amnesic agents, such as electroconvulsive shock; and drastic shifts in certain internal physiological states of the animal. Forgetting caused by these sources may be alleviated by either of two procedures, which may be characterized generally as providing retrieval cues. The manipulation of events so as to make the elements noticed by the subject during the retention test more similar to those noticed during original learning is the first technique for alleviating forgetting; this may be contrasted with the forgetting that would occur if the context of testing were permitted to deviate from the context of learning either through explicit experimental manipulations or through contextual changes, which may regularly occur under normal circumstances (see Spear, 1973). We know that a variety of factors, when changed between learning and testing, may impair retention: events associated with specific points in the circadian cycle (Stroebel, 1967; Holloway & Wansley, 1973); extent of food or water deprivation (e.g., Bailey, 1955; Levine, 1953); internal chemical activity of endogenous origin, including perhaps hormones and neurotransmitters (e.g., Spear, Klein, & Riley, 1971; Gray, 1975); a variety of chemicals of exogenous origin (Bliss, 1973; Overton, 1966, 1972); and common external environmental features, such as ambient illumination and specific physical features of a training apparatus (Chiszar & Spear, 1969; Zentall, 1970).

The other technique for alleviating forgetting in animals is termed "prior cueing." In this case, prior to the retention test the animals are exposed to a portion of the events believed to have been noticed during learning. The class of events most widely used for prior cueing has been the reinforcer–most commonly footshock in the case of aversive conditioning and food in the case of appetitive conditioning. We shall consider possible mechanisms and alternative interpretations of this effect later. For now, it is sufficient simply to note that presentation of the reinforcer prior to testing has been found to alleviate forgetting attributable to long or intermediate-length retention intervals, acquired conflicting memories, and a variety of amnesic treatments (Miller & Springer, 1973; Spear, 1973).

CONCEPTUAL FRAMEWORK

The conceptual framework within which we are working is broad, tentative, and flexible. Having elaborated elsewhere on this framework and its advantages and disadvantages (Spear, 1971, 1973, 1976), we shall restrict mention here to three of its features. First, a memory is a multidimensional representation consisting of a conglomeration of memory attributes that represent events noticed by an organism during learning. Second, retrieval of that memory occurs when the organism notices a sufficient number or kind of stimuli sufficiently similar to those represented as attributes of the memory. Third, a retrieved memory, whether or not manifested in overt behavior, is vulnerable to events that may modify the probability of retrieving that memory and, under certain circumstances when these events involve physiological insult, may modify the structure (the particular attribute components) of that memory. When retrieved, a memory may have characteristics (for example, vulnerability) similar to those found immediately after acquisition (for example, Lewis & Bregman, 1972; Lewis, Bregman, & Mahan, 1972). Implicit throughout is our optimistic view that a common mechanism may underlie all sources of forgetting, whatever their current nominal or operational distinctions. A corollary is that the mechanism underlying the alleviation of this forgetting may be common to all of the examples cited here.

We may now consider a series of experiments that we have designed and carried out between 1970 and 1975 in order to study the characteristics of this retrieval process and simultaneously to investigate certain features of the ontogeny of memory processing.

GENERAL PROCEDURE AND INITIAL TESTS OF RETENTION

The great majority of the experiments discussed here employ a procedure that has been described as the "hurdle-jumping, escape from fear task" (McAllister, McAllister, Brooks, & Goldman, 1972; for a thorough review of research using this procedure, see McAllister & McAllister, 1971). Fundamentally, the task involves Pavlovian conditioning procedures but measures instrumental responding to assess the influence of this conditioning. The specific procedure is quite simple. The apparatus is a rectangular box with two separate compartments in which, prior to training, each rat is given familiarization time, 7.5 min in the white compartment, which always has a grid floor, and 7.5 min in the black compartment, which typically has a wooden floor. A partition in the center of the apparatus separates the compartments when raised and, when lowered, forms a low hurdle that the rat must go over in order to leave the white compartment and enter the black compartment.

For conditioning in our experiments, the rat typically receives 30 pairings of a flashing light (the CS) and 1.0 mA footshock (the UCS) while in the white compartment. The CS has a duration of 7 sec and the UCS occurs throughout the final 2 sec of the CS. During the conditioning, the rat is unable to escape from either the CS or UCS. The rat is removed from the conditioning compartment between pairings and kept in a holding cage for 30 sec, just as between testing trials. For testing, the animal is placed in the white compartment, the partition separating the compartments is lowered (forming a hurdle), and the CS is initiated. With the barrier lowered, the rat has the opportunity to cross the hurdle and thereby escape the white compartment, simultaneously terminating the CS. A point worth emphasizing is that footshock never is delivered during testing. The index of the conditioned aversion to the white compartment and the CS is the progressive increase in the speed with which the animal crosses the hurdle and enters the black compartment.

The "hurdle-jumping escape from fear" procedure has six advantages of particular value for our purposes:

1. In general, Pavlovian conditioning procedures provide precise experimental control over the characteristics of the CS and UCS and their temporal relationship. At the same time, these procedures, unlike those of instrumental conditioning, eliminate the need to consider a memory attribute representing the instrumental response; such an attribute may have special and unique importance for animals (see Spear, 1973, 1976).

2. This procedure provides a discrete stimulus other than the CS, that is, the white compartment itself, which also acquires measurable control over behavior but may be acquired or forgotten independently of the CS.

3. The response requirements are minimal; none during conditioning and only running and slight jumping during testing. These responses are performed readily by rats of various ages and stages of muscular maturity.

4. The UCS is not required as part of the retention test. This permits a clearer evaluation of the capacity of the UCS to alleviate forgetting when it is presented prior to the retention test. If it were necessary to present the UCS during each testing trial, as in a relearning procedure, the relative influence of the UCS presented prior to testing obviously would diminish after the first test trial, that is, on the occurrence of the UCS during testing.

5. The consequences of this conditioning procedure, and many of the parameters that determine the behavioral consequences, are quite well understood as a result of the considerable amount of research previously conducted with this and analogous procedures (see McAllister & McAllister, 1971).

6. This procedure yields relatively rapid and thorough forgetting by rats of all ages, as we shall see from the following basic retention functions we have derived with this procedure.

Retention over Long intervals as a Function of Age during Conditioning

Our first task was to establish an estimate of the extent of forgetting that we could expect over a long retention interval. Accordingly, we simply conditioned groups of rats, returned them to their home cages for intervals of 1 day or 28 days, and then tested retention. We compared the retention of adults (60–100 days old) and weanlings (weaned at the age of 21 days and conditioned when 22–24 days old) because long-term retention after conditioning at these ages is known to differ under some circumstances (Campbell & Spear, 1972). At the same time these animals were conditioned, groups of animals with corresponding ages were treated identically in terms of familiarization with the apparatus but were given no conditioning trials and then were given a "retention" test after either 1 or 28 days. Such control conditions are usually only implied in studies of retention, but we have felt it important to include them to control unconditioned effects. Finally, an intermediate-length retention interval of 14 days was included for only the weanling animals to investigate whether their forgetting might be nominally complete long before 28 days. The results are shown in Fig. 6.1. In Fig. 6.1, and in subsequent figures, the term "memory component" is used to indicate the difference in retention performance between animals given the conditioning procedure (trained or conditioned) and animals not conditioned but otherwise treated the same (not trained or not conditioned). In other words, "memory component" simply indexes the behavioral consequence of conditioning after unconditioned effects, for the most part, have been subtracted.

It is clear from Fig. 6.1 that a good deal of forgetting occurs over 28 days with this task. When subjects are tested 1 day after conditioning, the consequences are documented thoroughly (see McAllister & McAllister, 1971), including a gradual increase in the speed with which the animals cross the hurdle into the dark compartment as the contingency between this response and termination of exposure to the white compartment and the CS is learned. Twenty-eight days later, this pattern is scarcely evident for either adults or weanlings. Although there is some numerical indication for a greater memory component in the adults, in neither case (adults or weanlings) did performance differences between animals given conditioning and those not given conditioning trials attain statistical significance. Moreover, the interaction between the age variable and conditioned vs not conditioned was not statistically significant with the 28-day interval (nor with the 1-day interval). Therefore, one cannot conclude that the weanlings forget more rapidly than the adults. However, any other conclusion is precluded until more thorough tests are applied because of the differing baseline of unconditioned performance with the 28-day interval.

In anticipation of a subsequent discussion, it is useful for us to present at this time a description of the retention functions found for still younger animals, infants 16–17 days old at the time of conditioning. A physical problem is present in the testing of animals at this age. Their small size and ambulatory deficiencies yield performance effects that preclude retention estimates if testing is conducted in the adult-sized apparatus. Therefore, for testing these younger animals' retention over intervals of 7 days or less, we have used an apparatus identical to that used for adults but scaled down in size, such that the volume is about 20% that of the adult-sized apparatus. Conditioning and testing in this apparatus works quite well for animals younger than 30 days; but by the time a rat is 44 days old, the age at which a 28-day retention test is given to animals trained as infants, they simply are too big to test in this small apparatus. Therefore, for retention intervals longer than 7 days, infants are conditioned and tested in the large apparatus, and for shorter retention intervals, they are conditioned and tested in the small apparatus. When a 7-day retention interval is used, we typically have replicated the experiment in both the small and large apparatus and have found no deviation in our basic results.

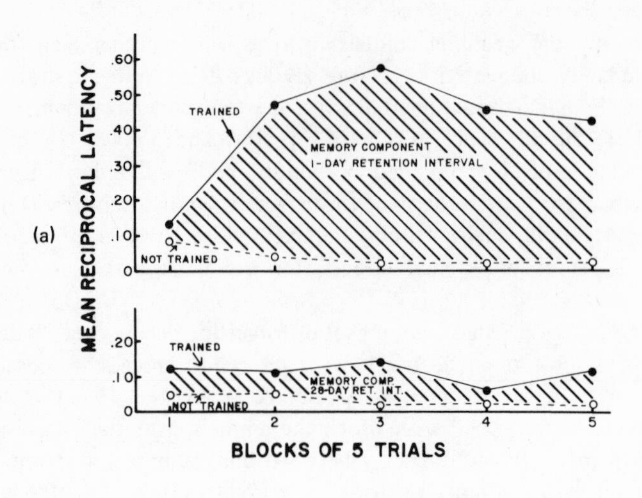

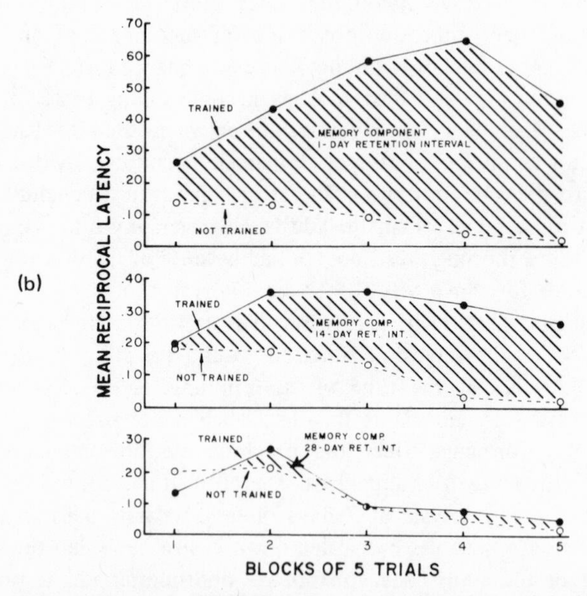

FIG. 6.1 (a) Retention in the hurdle-jumping, escape from fear task is shown for adult rats (80–100 days of age) on tests that followed original conditioning trials by either 1 day or 28 days. Effective conditioning is indicated during test trials by increasing jumping speeds (reciprocal latencies), which occur as the animal learns to respond in accord with the contingency between hurdle jumping and escape from those stimuli which had acquired aversive properties through conditioning. In each figure, the upper line represents the test performance of animals that had received the conditioning trials (trained) and the lower line (not trained) represents the performance of comparable animals that received equal handling but no conditioning trials. (b) This figure shows the retention by weanling rats (22–26 days of age during conditioning) when tested 1, 14, or 28 days following original conditioning. The details of the figure are the same as those of (a).

The course of forgetting in infant rats is shown in the lower half of Fig. 6.2 For purposes of comparison, the course of forgetting for weanling rats is shown in the upper half of Fig. 6.2. In this figure, the measures of retention after 3 min, 1 day, and 7 days were obtained by employing the smaller apparatus for conditioning and testing of both the infants and the weanlings. For measures after the longer intervals (14 or 28 days), the larger apparatus was used in conditioning and testing for both infants and weanlings. The most conspicuous result is the infants' rapid forgetting over 24 hr. The extent to which this may be caused by a lower degree of learning is unclear. Incidentally, performance after 3 min does reflect actual conditioning of the infants and is not an artifact of unconditioned inflation of activity or other consequence of the series of foot-shocks that terminated 3 min earlier: infants given an identical distribution of 30 footshocks in another apparatus, like those given no footshocks at all, do not learn to jump over the hurdle with the speed and persistence shown by animals that had experienced the conditioning procedure. A final point of interest,

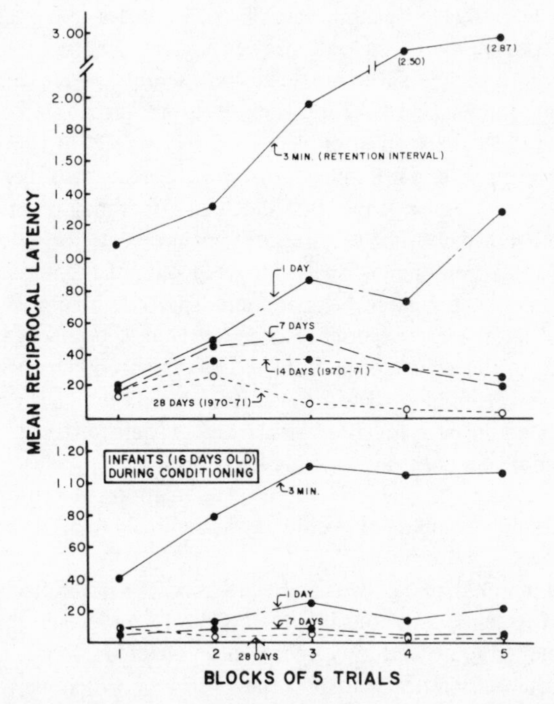

FIG. 6.2 Retention in the hurdle-jumping, escape from fear task is shown for weanling and infant (16 days of age) rats. All animals tested 3 min, 1 day, or 7 days following conditioning were trained and tested in the apparatus scaled to size for immature rats. Weanling animals (upper figure) given the 14- or 28-day retention intervals and infant animals (lower figure) given the 28-day retention interval were trained and tested in the adult-sized apparatus.

which is discussed later, is that retention by infants after 7 days is as poor as that of infants tested after a 28-day retention interval, at least insofar as our measurement permits an estimate.

THE REINSTATEMENT PARADIGM

A suggestion by Campbell and Jaynes (1966) has led to a procedural paradigm for alleviating forgetting, termed "reinstatement." The basic idea of Campbell and Jaynes is that early learning by immature organisms is manifested in later behavior to the extent that the animal has been reexposed in the interim to fractional learning experiences of the same nature as the early learning. In other words, a memory representing learning in early childhood is not forgotten if the individual occasionally becomes exposed to further learning trials even though by themselves–without the advantage of the original learning–these few scattered learning trials have negligible influence on behavior. As a specific example, Campbell and Jaynes gave weanling rats 30 pairings of a specific stimulus and a footshock and found that these rats showed a great deal of forgetting 28 days later. This forgetting was alleviated to a considerable extent if three discrete pairings of that stimulus and the footshock occurred, distributed throughout the retention interval. Animals given only these three widely distributed conditioning trials, but no prior conditioning, showed little evidence of conditioning that could be attributed to these three trials. Such an effect may provide a resolution to the paradox in which poor long-term retention exists for specific memories acquired in infancy (Campbell & Spear, 1972) in spite of the importance of early experience generally for later behavior and survival. Memories acquired for specific episodes of early experience apparently do persist to influence later behavior when, in the course of maturation, the episode recurs periodically as in the reinstatement paradigm. The mechanism underlying such conservation of early memories could be especially pervasive if the reinstatement effect did not require recurrence of a complete training-trial episode and, instead, some component of the complete trial were sufficient to maintain retention. Surely, the occurrence of some component would be more likely than a replication of a complete episode.

We wish to emphasize at this point a critical distinction between those reactivation treatments consisting of a complete reconditioning trial and those that do not, including instead only one or more selected components of original conditioning: the former clearly involves new learning, which may summate with that of original conditioning, but the latter probably does not. Of course, the question of whether any particular reactivation treatment yields learning that is consistent with original conditioning must be an empirical matter. Yet, we should recognize that whereas any reactivation treatment may, with a certain inevitability, induce new learning of some sort, exposure to isolated components of conditioning may induce learning that is at least as likely to subtract from that of original conditioning as to add to it. This is most obvious in the case of an explicit CS; presentation of only the CS, without a following UCS, constitutes the conventional extinction treatment. However, isolated presentation of contextual components of conditioning also may act as extinction treatments, as may isolated presentation of the UCS (see, e.g., Gibbon, Berryman, & Thompson, 1974). We see no inconsistency in attributing to an event both extinction and reactivation consequences; indeed the latter may be a prerequisite of the former. Evidence is available from diverse sources to indicate conditions under which extinction treatments may have reactivation consequences (Spear, 1976).

Because of the analytical potential of the reinstatement paradigm, we (together with Clifford Ott) initiated further tests to investigate which, if any, components of training are sufficient to alleviate forgetting with this procedure (also see Silvestri, Rohrbaugh, & Riccio, 1970). Of particular interest were two loosely drawn hypotheses. The first hypothesis was that reinstatement might be uniquely beneficial to memories acquired prior to maturity. Because physiological maturation and perceptual learning possibly might cause changes in a maturing animal's perceptual encoding of environmental events, periodic reexposure to some elements of original learning would permit the maturing animal to accommodate the encoding of these events to his changes in perception. Such "perceptual accommodation" seems a reasonable means through which a maturing animal may continue to respond appropriately to particular stimuli in spite of changes in the animal's perception of these stimuli [an analogous construct termed "accommodation," although not restricted to perception per se, has been employed in the interpretation of memory processing in children (Piaget & Inhelder, 1973)]. Without perceptual accommodation, changes in an animal's perception of the conditioning situation could lead to a deficit in retention performance. It seems reasonable also to expect that following learning, maturing rats would have a greater need for perceptual accommodation than adults. If these guesses about perceptual accommodation are correct, we can expect alleviation of forgetting through reinstatement procedures—which provide periodic reexposure to some elements of original learning—to be greater for animals trained as weanlings than for those trained as adults.

The second hypothesis focused on the notion that, depending on age, animals deal differently with various memory attributes. One possibility was that the typically greater forgetting by younger animals could be traced to their greater forgetting of certain memory attributes representing particular events, rather than to uniformly greater forgetting of all attributes. In other words, the memory attributes representing certain events, such as aversive stimuli, may be forgotten more rapidly by weanlings than by adults, but retention of other memory attributes may be equal for weanlings and adults. Another possibility was that the memory attributes of weanlings and adults simply differed, that is, immature and mature rats learned different things as a consequence of conditioning. Both of these possibilities would predict a retention difference between weanlings and adults when the reinstatement procedure included certain components of the conditioning situation but the absence of a difference when certain other elements were included.

Using weanling and adult rats, our initial experiment compared the effects of reexposing the rats during a 28-day retention interval to one of four events that had been associated with original conditioning: presentation of a complete conditioning trial (a CS–UCS pairing); reexposure to only the white compartment of the apparatus; simultaneous reexposure to both the white compartment of the apparatus and the CS; and reexposure to only a footshock delivered in a different apparatus. Eight separate groups of rats, four of weanling age (22–26 days) and four of adult age (60–100 days), received 30 conditioning trials (footshock paired with a flashing light in the white compartment) with the procedure described earlier and were tested for retention 28 days later. Each animal received just one of the four alternative reinstatement treatments, presented on six occasions distributed at approximately equal intervals throughout the 28 days, the last treatment occurring 4 days prior to the retention test. An additional eight groups of rats, four of weanling age and four of adults, were familiarized with the apparatus but were not given conditioning trials. Each of these control groups was given one of the four reinstatement procedures and a "retention" test, exactly as the conditioned animals. In passing, we should add that the reinstatement effect for those conditioned animals given the complete conditioning trial is uninterpretable because so much learning occurred as a consequence of the six reinstatement treatments alone, as shown by the unconditioned control subjects; the other reinstatement treatments predictably had negligible effect in the absence of original conditioning.

The animals exposed to only the white compartment as the reinstatement treatment did show a slight reduction in forgetting as a consequence (a result probably analogous to that of Silvestri *et al.,* 1970, Experiments 1 and 2), whereas those animals given exposure to both the white compartment and the CS performed no differently than would have been expected without a reinstatement procedure. However, the results of greatest interest for this chapter are those of animals the reinstatement treatments of which consisted of exposure to the UCS only.

Forgetting was alleviated markedly among animals exposed to only the UCS throughout the retention interval, as shown in Fig. 6.3. When we compare the top and middle panels of this figure, we can see that the reinstatement effect for weanlings tends to be less than that for adults. Moreover, for still younger animals (infants, 16 days of age, treated in the same way in a later experiment) this UCS reinstatement treatment failed entirely to alleviate forgetting. We shall return shortly to a consideration of the ontogenetic relationship indicated in Fig. 6.3. For now, we shall concentrate on further analysis of the basic reinstatement effect found with adult animals, an effect that has surprised us with its potency.

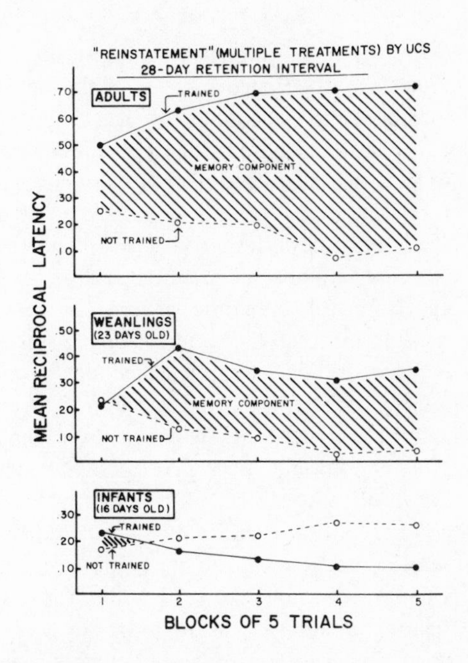

FIG. 6.3 The effect of presenting the UCS as a reactivation treatment within the reinstatement paradigm is shown for adult (top), weanling (middle), and infant (bottom) animals tested 28 days following conditioning (trained) or following equivalent handling but no actual conditioning trials (not trained). All animals received presentations of the UCS in a neutral apparatus on six equally distributed occasions throughout the 28-day interval. The higher the reciprocal latencies, the better the retention performance.

ALLEVIATION OF FORGETTING BY A SINGLE REACTIVATION TREATMENT: THE REACTIVATION PARADIGM

Two immediate questions arose from the success of the UCS reinstatement treatment. In relation to our pretheoretical framework, one question was cautiously pessimistic, the other wildly optimistic. We first were concerned with contamination by whatever learning might have occurred as a consequence of presenting the UCS during the retention interval. Conceivably, new learning generated by the six interpolated UCSs in our experiment above merely summates with original conditioning rather than causes perceptual accommodation or periodical retrieval and consequential maintenance of the memory representing original conditioning. However, if such summation is the basic effect of re-presenting the UCS, then it is an extraordinary and interesting instance of new learning in its own right: we have found in our reinstatement experiment that conditioned rats given the UCS reinstatement treatment show retention far better than that of animals overtrained originally by an equivalent number of complete CS–UCS pairings and is equal to that of animals whose reinstatement treatment has consisted of complete conditioning trials. The implication is that if the UCS reinstatement effect results from new learning, then the learning caused by distributed presentations of the UCS alone is far greater than that caused by massed presentations of complete conditioning trials and equal to that caused by distributed presentations of complete conditioning trials. This seems unlikely. Moreover, subsequent control experiments (which we shall discuss shortly) encourage us in the belief that the effect is primarily one of memory retrieval and not new learning.

The second, more optimistic, question concerned the extent of the influence of the UCS on retention. For example, would presentation of the UCS alleviate forgetting even though it was not presented until just prior to the retention test? If so, an interpretation in terms of new learning would be less likely with this decreased number of presentations of the UCS. Furthermore, if forgetting were alleviated for immature as well as mature animals, an explanation of the reinstatement effect in terms of perceptual accommodation would be unlikely because the UCS treatments would be given only at the end of the retention interval rather than being distributed throughout the interval in accord with the animals' maturation. Therefore, in a small study, subjects were tested 28 days after conditioning and were presented with the UCS alone 24 hr prior to the test. For reasons primarily superstitious, this "reactivation treatment" actually consisted of two UCS presentations separated by a 5-min interval in which the subject was returned to its home cage (subsequently we learned that a single UCS presentation was sufficient). The 24-hr interval between the reactivation treatment and the testing was chosen for more sound reasons–to minimize unconditioned consequences of this treatment that might occur if it were presented only a few minutes before testing. We were pleasantly surprised to find that alleviation of forgetting with presentation of the UCS limited to one occasion 24 hr prior to testing apparently was as great as that found with the several distributed presentations of the UCS given with the reinstatement paradigm. Hereafter, we shall refer to the former procedure as the "reactivation paradigm," as opposed to the latter procedure, termed the "reinstatement paradigm."

An obvious need at this point was thorough confirmation of the influence of the single reactivation treatment given 24 hr before the retention test. We conducted several experiments, each of which included tests of this basic effect but incorporated also tests of various other issues (which may be ignored here.) That the basic effect of the single reactivation was confirmed is shown in Fig. 6.4, a composite which combines the results of key conditions that were replicated among three experiments. This figure clearly illustrates a striking benefit to the animal's retention performance from simple exposure to the UCS 24 hr prior to testing, provided the animal previously was conditioned with pairings of the CS and UCS (trained, UCS reactivation). Figure 4 also shows that neither conditioning alone (trained, no reactivation) nor the UCS reactivation treatment alone (not trained, UCS reactivation) were sufficient to produce the effect.

To further illustrate the robustness of this reactivation effect, we may cite the results of one experiment. The specific details may be omitted, except to note that each of five reactivation treatments tested included presentation of the UCS

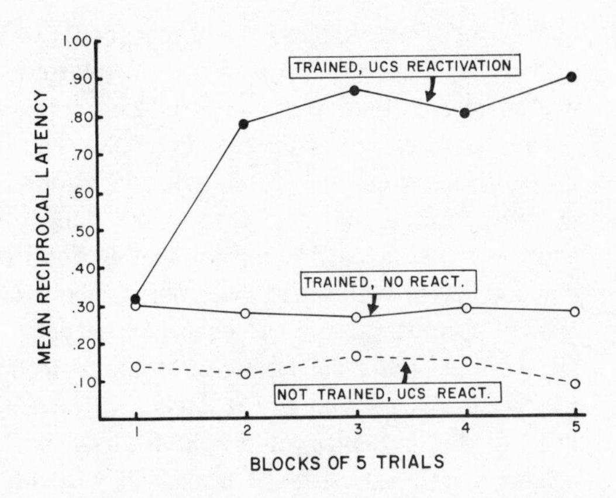

FIG. 6.4 Retention performance in the hurdle-jumping, escape from fear task is shown for adult animals tested 28 days after either conditioning (trained) or equivalent handling (not trained) and either presented the UCS 24 hr prior to testing (UCS reactivation) or not.

24 hr before the retention test. One treatment included only the UCS, but for each of the others the CS preceded the UCS by a certain interval. Each treatment markedly enhanced retention performance, with the effect of the UCS alone being so potent as to overshadow that of all other combinations of the UCS and CS, even the influence of presenting the CS and UCS exactly as in a complete conditioning trial. Having established the magnitude of the effect, we became concerned about artifactitious consequences of the UCS that might have inflated retention performance independent of memory retrieval.

ANALYSIS OF THE REACTIVATION EFFECTS CAUSED BY PRESENTING THE UCS ALONE

We were concerned that the use of a wooden floor in the black (safe) compartment might have exaggerated the consequences of the learning that resulted from the rat receiving the footshock UCS on a grid floor during the reactivation treatment. Perhaps this experience caused the rat to learn that all grid floors are bad and so increased its preference for the wood floor in comparison with the grid floor in the white compartment, which would appear to us as enhanced retention performance. A major flaw in this interpretation is that animals given the UCS reactivation treatment but not previously conditioned behave no differently than animals given no reactivation treatment at all (see Fig. 6.4), that is, they seem to have learned little or nothing about the aversive characteristics of grid floors as a consequence of only the UCS presentation. Simple exposure to the UCS was not sufficient to produce the reactivation effect; clearly, prior conditioning also was necessary. However, there remained a remote possibility that the distribution of the "learning about grid floors" trials—an interval of 27 days between conditioning and the reactivation treatment—somehow might account for the dramatic improvement in memory retention performance (which again would be a most interesting effect in itself, if true).

We conducted three sets of experiments to investigate the possibility of new learning during the UCS reactivation treatment. The obvious initial experiment was to test animals with the grid floor present in both compartments of the apparatus, thus removing the difference in floor texture that otherwise might distinguish "dangerous" and "safe" locations for the rat and enhance the value of new learning that grid floors are "dangerous." The UCS reactivation treatment was presented 24 hr prior to the retention test to two groups of adult rats that 28 days earlier had received either 30 or 120 conditioning trials. Two control groups of rats also received either 30 or 120 conditioning trials but received no additional treatment prior to testing 28 days later. The results (see Fig. 6.5) indicate that the reactivation treatment clearly alleviates forgetting in spite of the presence of a grid floor in both compartments. The apparently greater reactivation effect for animals with a higher degree of original learning is significant to our later discussion but is irrelevant to our present point. Somewhat troublesome was the fact that the net alleviation of forgetting caused by the reactivation treatment was somewhat less than we previously had obtained. This was not especially surprising in itself, but the source of the lessened effect was unexpected: the primary factor appeared to be the better retention performance of animals given no reactivation treatment at all. The implication was that

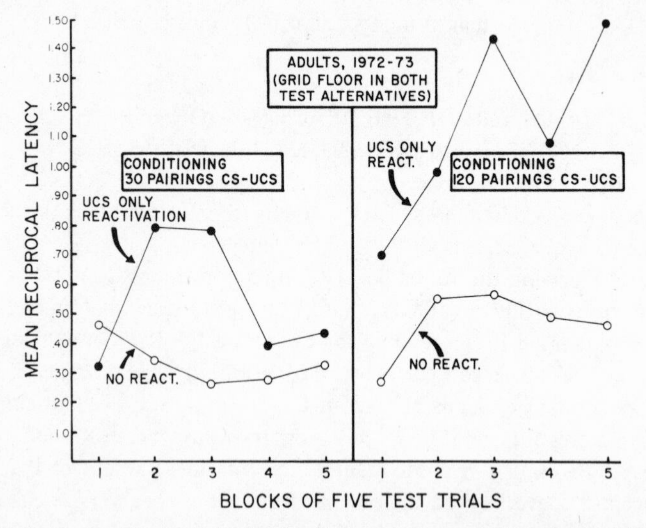

FIG. 6.5 Retention performance in the hurdle-jumping, escape from fear task is shown for adult rats tested 28 days after original conditioning. Animals in the left panel had received 30 pairings of the CS and UCS and animals in the right panel had received 120 pairings. Half of each of these sets of animals were exposed to the UCS in a neutral apparatus 24 hr prior to the retention test (UCS only reactivation) and the remaining animals were not. For this test, the floor of the "safe" compartment, to which the rats escaped during testing, was a grid floor identical to those of the "dangerous" compartment from which they escaped and of the neutral apparatus in which the UCS reactivation treatment was administered.

rats rather preferred to stand on a grid floor than on a wooden floor (McAllister & McAllister, 1971, also have reported this) and therefore they more readily jumped the hurdle into the black compartment when the grid floor was present, independent of any prior conditioning or reactivation treatment. We nevertheless may conclude from the major results of this experiment that presentation of the UCS 24 hr before testing substantially alleviates forgetting, whether or not the texture of the floor associated with the reactivation treatment differs from that of the "safe" compartment of the testing apparatus.

As a further test for a contaminating influence of new learning mediated by the kind of floor on which the UCS reactivation treatment was applied, an additional experiment was conducted in which the wooden floor in the apparatus did not differentiate the dangerous and safe compartments. This time a slightly different source of forgetting was employed, with a procedure designed to minimize the stimulus value of the grid floor per se. For several days prior to conditioning and throughout the 7-day retention interval, animals in this experiment were housed in group cages that included a replica of the conditioning apparatus, with the grid floor but minus the center door and end panels so that the animals could traverse easily through the apparatus. As shown in Fig. 6.6, the effect of housing with the simulated apparatus present was to decrease retention performance in general, perhaps because of poorer original conditioning. However, forgetting was significantly alleviated for subjects that received the

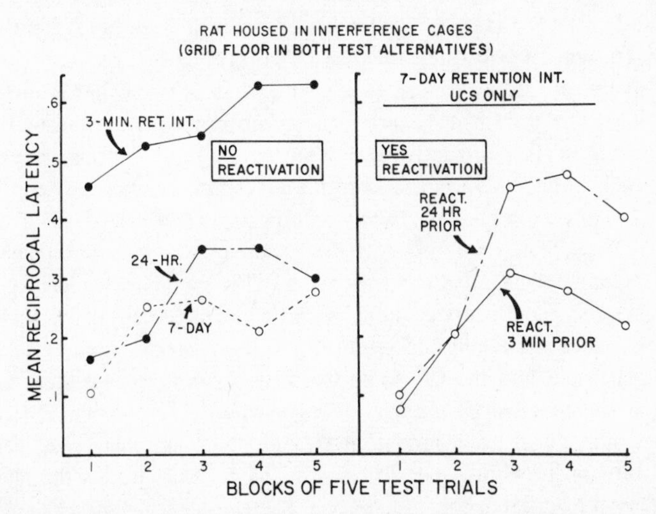

FIG. 6.6 Retention performance is shown for adult rats housed with replicas of the experimental apparatus and tested with identical grid floors in the "safe" and "dangerous" compartments of the apparatus and in the neutral apparatus in which the UCS reactivation was administered. The left panel shows retention following differing intervals, during which no reactivation treatment was given. The right panel shows retention following a 7-day interval, with the UCS reactivation treatment given either 24 hr or 3 min prior to testing.

UCS reactivation treatment 24 hr prior to the retention test. Surprisingly, corresponding alleviation of forgetting was not obtained when the UCS reactivation treatment preceded the test by only 3 min, presumably because persisting unconditioned consequences of the UCS disrupted, rather than facilitated, speed of jumping from the white compartment to the black compartment.

A third set of experiments (which need not be elaborated here) also indicated that the UCS reactivation effect could not be explained in terms of the similarity between the grid floors associated with reactivation and with the "dangerous" compartment at testing. In this experiment, which involved immature rats, the UCS reactivation treatment was administered on a grid that was made of flat bars close together, unlike the round, relatively well-separated bars of the testing apparatus grid; nevertheless, the reactivation effect occurred as robustly as ever.

A Test for Activity Artifacts

One might wonder whether the presentation of noncontingent footshock 24 hr prior to the retention test inflated the animals' general activity levels, resulting in faster speed of traversing from the white to the black compartment. Such an artifact would require a very complex mechanism, because animals given the reactivation treatment without prior conditioning clearly did not show this kind of "activity" effect. If enhanced activity indeed did inflate retention scores, therefore, it must be dependent somehow on the interaction between prior training and the reactivation treatment. To test this remote possibility, we altered the nature of the retention test. Two groups of rats were conditioned (again, with the procedure described earlier), whereas two other groups received no conditioning trials; 28 days later, one conditioned group and one unconditioned group were given a reactivation treatment. All four groups were tested for retention 24 hr later: the animals were placed in the black (safe) compartment and their latency to enter the white compartment with the CS present was measured. Of particular interest was the group of animals that received both conditioning and the reactivation treatment. With such a test, speed of responding by previously conditioned animals would be expected to be lower if the reactivation treatment facilitated retention of the aversive characteristics of the white compartment and the CS. In contrast, one would expect these previously conditioned and reactivated animals to respond more rapidly than either animals conditioned and given no reactivation treatment or animals given reactivation treatment but not previously conditioned, if the consequence of the reactivation treatment were to inflate general activity specifically among animals conditioned 28 days earlier. In fact, speed of jumping into the "dangerous" compartment with the CS present was quite slow in each condition, and those animals given both conditioning and the reactivation treatment tended to be slowest of all. We have concluded that the reactivation effect is not an artifact of activity.

A Further Note on the Possible Unconditioned Influences of the UCS Reactivation Treatment

Perhaps the UCS (footshock) 24 hr prior to the retention test induces in the animal a general state of "fear," in the sense that those physiological processes activated when an animal is exposed to pain, or threat of pain, may be activated by the UCS and persist throughout the retention test. Such persistence might modify the contemporary context of the retention test in such a way as to make the animal's internal physiological state more similar during testing to that during conditioning than otherwise would occur in the absence of footshock. The assumption is that the persistence of the unconditioned consequences of the footshock continues over a period of 24 hr. We have mentioned that this seems unlikely, but it must remain a possibility. It certainly would be reasonable to expect such effects to dissipate over time. If this hypothesis were correct, therefore, a UCS presented only a few minutes prior to the retention test should have even more influence on retention than a UCS presented 24 hr prior to the test. In Fig. 6.6, we have the results of a test of this hypothesis. Clearly, the UCS presented 3 min prior to testing yielded less, not more, facilitation of retention than the UCS presented 24 hr earlier (this difference was not statistically significant, however).

ONTOGENETIC ANALYSIS OF THE INFLUENCE OF THE UCS REACTIVATION TREATMENT

Our analysis of the reactivation effect in adult animals was paralleled in experiments with younger ones. We already have seen in Figure 6.3 the relative influence on retention of presenting the UCS on six equally distributed occasions throughout the retention interval within the reinstatement paradigm. We have also gathered data that permit the same ontogenetic comparisons within the reactivation paradigm; in this paradigm, no treatments are given animals throughout the retention interval until 24 hr prior to the retention test. We conditioned rats as infants (16 days old), weanlings (22–24 days old), or adults (60–80 days old) and tested retention 28 days later. This test was preceded 24 hr earlier by presentation of the UCS in a neutral apparatus. The estimate of the memory component was derived by comparing performance of these rats with the performance of comparable rats of each age treated identically except that they were not conditioned. The general pattern of the results, shown in Fig. 6.7, is strikingly similar to the pattern of results for the reinstatement paradigm. In both cases, forgetting was alleviated significantly for both adults and weanlings, with a tendency toward less of an effect for the weanlings. Also in both cases, forgetting was not at all alleviated in animals trained as infants. To summarize,

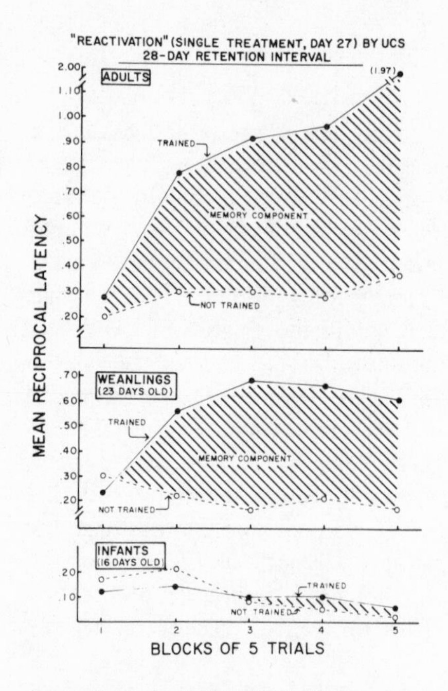

FIG. 6.7 Retention performance in the hurdle-jumping escape from fear task is shown for adult (top), weanling (middle), and infant (bottom) animals previously given 30 conditioning trials (trained) or merely given equivalent handling (not trained). Conditioning trials were given at either of three ages (infant, weanling, or adult) and retention was tested 28 days later. All animals in this figure were exposed to the UCS, presented in a neutral apparatus, 24 hr prior to the retention test.

presentation of the UCS during the retention interval–whether on six occasions distributed throughout the interval or on two contiguous occasions (separated by 5 min) 24 hr prior to the retention test–served to maintain the memory component of earlier conditioning rather well in adults, somewhat less well in weanlings, and not at all in infants, when tested after a 28-day retention interval. We now comment on where these results may lead us.

Perceptual Accommodation

These data indicate that perceptual accommodation is not an important determinant of alleviated forgetting within the present reinstatement procedure. The perceptual accommodation notion would predict greater benefit from the reinstatement procedure for younger, maturing animals, but there was no indication of this relationship within this paradigm; indeed, the opposite relationship prevailed. Moreover, the effect of age with the reinstatement paradigm was identical to that found with the reactivation paradigm, and in the latter there was no opportunity for perceptual accommodation because the UCS was not presented until just before the retention test. Furthermore, there is no indication, for rats at any age, that alleviation of forgetting is greater with the reinstatement paradigm than with the reactivation paradigm; if progressive accommodation to perceptual change is the major benefit from reexposure to events of conditioning, one expects instead relatively greater alleviation of forgetting with the reinstatement paradigm because it provides opportunity for perceptual accommodation that the reactivation paradigm does not. It would appear that the effect of, or opportunity for, perceptual accommodation was insignificant.

We should note, however, that comparison of these particular applications of the reactivation and reinstatement paradigms involves three confounded factors of potential importance. With the reinstatement paradigm the UCS was presented on six, widely distributed occasions, whereas with the reactivation paradigm the UCS was presented on two, relatively contiguous occasions. In addition, the interval between the last presentation of the UCS and the occurrence of the retention test was 4 days in the case of the reinstatement paradigm and 1 day in the case of the reactivation paradigm. Any of these three factors—number of UCS presentations, distribution of UCS presentations, or interval between the final UCS presentation and the test—might influence retention performance. Therefore, insofar as the data presented in Figs. 6.3 and 6.7 are concerned, perceptual accommodation seems to be a negligible factor, although the generality of this conclusion is unclear because of the limitations of our design and procedure. In a moment, we shall consider an experiment that has provided results derived from the reinstatement and reactivation paradigms that may be compared more meaningfully.

Interpretations Concerning Memory Storage and Retrieval

The results with the reactivation paradigm (Fig. 6.7) suggest that presentation of the UCS may have resulted in direct retrieval of the memory of the conditioning episode or direct retrieval of some major attributes of that memory. By "direct retrieval" we mean that accessibility of attributes of that memory achieved a state similar to that state occurring as an immediate consequence of conditioning. In a very gross sense, the memory components shown for the adults, weanlings, and infants in Fig. 6.7 are not substantially different from the memory components that should be found for adults, weanlings, and infants tested 24 hr after conditioning (see Figs. 6.1 and 6.2). In this respect, the memory component for adults in the reactivation paradigm seems inflated somewhat, especially in comparison with that of the weanlings; however, the test is not refined sufficiently to warrant speculation about this deviation. The really striking effect, of course, is with the animals conditioned as infants; for these animals, the consequences of reactivation are nil in comparison with the substantial effects found with the older animals. Animals in the infant group did perform at about the same level as would infants tested 24 hr after they were conditioned, displaying little or no retention (see Fig. 6.2); in contrast, the infant animals were 44 days old at the time of the retention test, and considerably more retention would be expected of an animal of this age if conditioning had occurred 24 hr earlier.

In a general sense, the reactivation paradigm may have failed to yield alleviated forgetting in the infants for either of two reasons. Perhaps infant rats simply do not have the equipment needed for storage of a memory that is to persist over long intervals. Likewise, perhaps memory storage in animals of this age becomes relatively inaccessible to the retrieval process after a long interval. Both alternatives have been mentioned by Campbell and Spear (1972) and may be stated in other words: first, there may be a storage deficit in infant rats in the sense that there is no memory available for retrieval after a 28-day interval; second, presentation of the UCS may fail to initiate retrieval of the memory in spite of the adequate storage, possibly because the consequences of the UCS are represented so differently in the 43-day-old rats compared to the 16-day-old rats (maturation may result in a variety of central and peripheral changes in the neurohormonal and sensory consequences of a footshock). This is rather like suggesting a change in the "schemata" through which the animal analyzes his world, but at a more concrete biological level than is implied typically when the term "schemata" is used to refer to cognitive functioning.

Toward a decision between these alternatives, we applied the reinstatement and reactivation paradigm to infant rats conditioned at 16 days of age and tested 7 days later, rather than 28 days later. The reinstatement and reactivation paradigms were alike in this experiment in including one UCS presentation to all animals 24 hr prior to testing; in addition, animals in the reinstatement condition received one footshock on Days 2, 4, and 6, considering Day 1 as the conditioning day and Day 8 as the day of testing. Independent groups of infant rats, although not conditioned, were given familiarization and either the reinstatement or the reactivation treatment prior to their "retention test." The results are shown in Fig. 6.8, together with the results of 16-day-old animals given the reactivation or the reinstatement paradigms and tested after a 28-day retention interval. It is apparent that forgetting after the 7-day interval has been alleviated substantially in these animals within both paradigms. This immediately dispels the possibility that memory storage is a very transient feature of infant learning and also indicates clearly that retrieval is possible for at least 6 days after original conditioning.

This experiment permits comparison of the relative effects of reactivation and reinstatement treatments on retention after a 7-day interval. The comparison is complicated by the unconditioned consequences found with the reinstatement paradigm–rapid responding by animals given four footshocks distributed over the retention interval but no conditioning trials. However, in spite of the absolute rise in test performance attributable to the unconditioned consequences of these multiple footshocks (an effect we do not yet understand), the net benefit to the retention of the conditioned animals tended to be greater within the reactivation paradigm. This may imply an extinction effect of the multiple footshocks, acting in opposition to the unconditioned effect with the reinstatement paradigm. However, our preliminary tests (with adult animals) have not

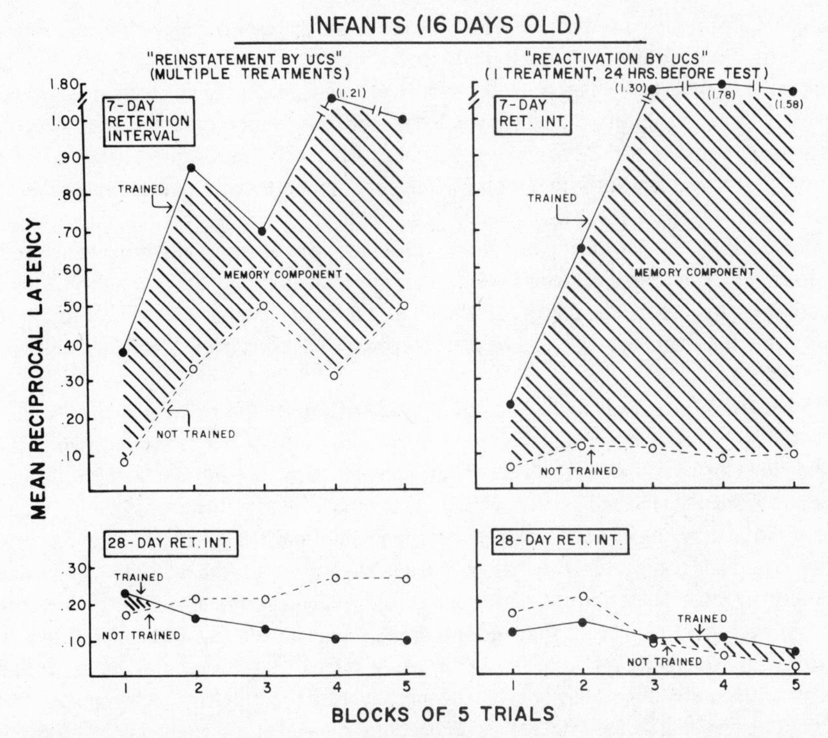

FIG. 6.8 Retention performance in the hurdle-jumping, escape from fear task is shown for infant rats given 30 conditioning trials (trained) or equivalent handling at 16 days of age. The retention test was administered after either 7 days (upper panels) or 28 days (lower panels). Animals represented in the left portion of the figure were presented the UCS reactivation treatment on several occasions throughout the retention interval; those represented in the right portion of the figure were presented the UCS reactivation treatment only once, 24 hr prior to the retention test.

confirmed this notion and further tests with infants are needed to develop our understanding of these apparent differences found between effects of the two paradigms.

These small differences aside, the evidence from both paradigms of excellent retention in infants 7 days after conditioning generates a number of questions. If the memory is retrievable 6 days after conditioning, why has the reinstatement procedure not been able to maintain the memory for subsequent retrieval over the 28-day retention interval, in view of the fact that the first presentation of the UCS has occurred less than 6 days after training? Apparently, the immediate consequences of presenting this UCS are not identical to those of the complete learning experience—the set of 30 conditioning trials (a point that warrants mention only within the context of the results described in this chapter). Another question concerns the magnitude of the memory component when a single UCS is presented 24 hr prior to the retention test: the retention exhibited appears far greater than that likely for an infant actually conditioned 24 hr earlier. Of course, by the time the reactivation treatment was given, the infants had become weanlings, and as we have seen, weanlings do exhibit a sizeable memory component 24 hr after training. Their absence of retention after a 28-day interval in both the reactivation and reinstatement paradigms makes it seem unlikely that the "weak" memory of the infant is converted after a 6-day interval into a memory characteristic of the weanling—a memory somehow "stronger" but apparent only with the proper retrieval aids. The implication, perhaps barely conceivable, is that such a "strength" characteristic of a memory may be independent of its absolute probability of retrieval (cf. Wickelgren, 1974).

Such questions lead no further than speculation at this point because we lack appropriate tests. However, we may offer preliminary data relevant to two other questions of equal interest to us. Both questions center around a critical portion of the results described above, which is summarized as follows: rats conditioned as infants show more rapid forgetting than rats conditioned at an older age and the magnitude of their forgetting is such that after either a 7-day or 28-day retention interval, retention performance by rats conditioned as infants does not differ significantly from that of an animal that never has been conditioned. However, forgetting by infants over a 7-day retention interval may be alleviated with either the reinstatement or the reactivation paradigm, but forgetting by infants after a 28-day interval is unaffected by either of these procedures. In contrast, forgetting after a 28-day interval is alleviated substantially with either paradigm for animals conditioned as weanlings or adults.

The first question is whether the effect obtained with the infant animals is unique to animals undergoing a rapid rate of physiological maturation during the retention interval. As a preliminary gross test of this question, we investigated forgetting and the consequences of reactivation treatment in very old animals, in the range of 14–16 months of age (which may be characterized as "elderly"). We found that when animals of this age are exposed to the same conditioning procedure as those in the experiments described above, forgetting is substantial after both 7 days and 28 days, in a quantity very much like the forgetting found in infants. This in itself is potentially of theoretical interest, provided we can attain some control over perception and degree of original learning, as we shall discuss shortly. Also like the infants, elderly animals exposed to the reactivation paradigm—presenting to them the UCS 24 hr prior to testing after a retention interval of 28 days—show no alleviation of their forgetting, as shown in Fig. 6.9. However, the consequences of reactivation 24 hr prior to testing after a 7-day retention interval were substantial: elderly animals, like the infants given the identical treatments, showed quite excellent retention. This clearly indicates that the effects found in animals conditioned as infants are not a unique consequence of the infant's rapid rate of maturation during the retention interval. Perhaps progressive physiological change in general is responsible for these effects,

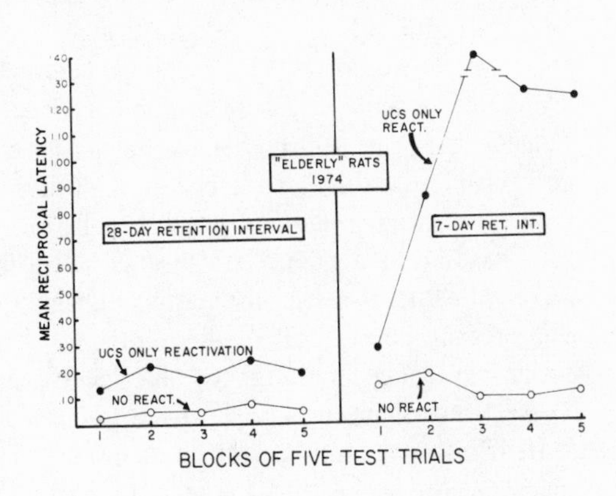

FIG. 6.9 Retention performance in the hurdle-jumping, escape from fear task is shown for "elderly" rats given 30 conditioning trials at 14–16 months of age. Animals in the left portion of the figure were tested after a 28-day retention interval and those in the right portion were tested after a 7-day interval. Half of the animals tested after each interval were presented the UCS reactivation treatment in a neutral apparatus 24 hr prior to testing (UCS only reactivation) and the remaining animals received no treatment between conditioning and testing (no reactivation).

growth in animals conditioned as infants, deterioration in elderly animals. However, it is not at all clear that the same processes cause these similar phenomena.

The major analytical impediment in all of these studies comparing retention after conditioning at different ages is the uncertainty concerning degree of original learning. In the experiments described above, the parameters of conditioning were held constant across ages. It seems quite likely that the effective conditioning (that is, degree of learning) resulting from these procedures is greatest in the adult and weanling animals and progressively less for animals younger or older. Furthermore, our data indicate that the consequences of reactivation for adults tend to be of greater magnitude when there is a higher degree of original learning, provided measurement artifacts, such as ceiling effects, are eliminated (see Fig. 6.5). The procedures were straightforward: conditioning proceeded exactly as in the experiments described above, except for the changes mentioned earlier, housing the rats with the replica of the apparatus and testing with grid floor in both alternatives. This served to decrease the overall level of conditioning and thus avoided ceiling effects in measurement. Also, half of the animals received 30 pairings of the CS and UCS, and half received 120 pairings. It is evident from these results that retention performance was greater for animals given the greater number of conditioning pairings and the consequence of the reactivation treatment was greater for animals given more conditioning trials. It is not certain that degree of original learning can account for the entire array of results described above for the infant and elderly animals, but we cannot know until this parameter is investigated properly. Our tests of this factor are not yet complete.

We may address, however, the more general question of whether degree of alleviation of forgetting by reactivation treatment depends on age at original learning. For certain instrumental learning tasks, the characteristics of original learning and of retention after short intervals are known to be identical in weanling and adult rats, even though after a long retention interval more forgetting is found for the weanlings (Campbell & Spear, 1972). One-way active avoidance is such a task (Campbell & Spear, 1972; for procedural details, see Feigley & Spear, 1970; Klein & Spear, 1969); using this task, we trained adult and weanling animals to a criterion of five consecutive avoidances and tested retention 28 days later by providing the animals with the opportunity to avoid for five trials in the complete absence of footshock. We tested the effects of four reactivation treatments, the nature of which are of no particular significance for our present point. Half of the adults and half of the weanlings did not receive active avoidance training but did receive an equivalent amount of handling. These control animals were given the reactivation treatments prior to their "retention test," thus permitting us to estimate of the memory component. The memory component was calculated (as above) by subtracting the retention score of animals trained on active avoidance from that of animals not trained but otherwise treated the same. In this case, the retention test score is specified in terms of the probability of an avoidance response (where an avoidance response is defined as a response with a latency less than 5 sec). The memory component after the 28-day retention interval was +.15 for adult rats if no reactivation treatment was given prior to testing; this score increased to +.70 if the most effective reactivation treatment (a treatment identical to the "warmup" treatment applied by Spear, Gordon, & Martin, 1973, Experiment 5) was given 24 hr prior to testing. For weanlings, the comparable scores were slightly negative in value, both for animals that received the reactivation treatment and for those that did not receive it, indicating considerable forgetting by weanlings and little or no alleviation of their forgetting when the reactivation treatment was administered. More generally, those reactivation treatments that did alleviate forgetting were effective only for animals trained as adults, not for those animals trained as weanlings. Again, this influence of age during conditioning on subsequent alleviation of forgetting occurred even though the number of trials required initially to learn this task did not differ between the adults and weanlings. This finding encourages our view that the ontogenetic differences in the results obtained with the reactivation and reinstatement paradigms may be caused by factors of broader theoretical interest than degree of original learning.

SUMMARY AND COMMENT

We have described our preliminary investigation of a particular reactivation treatment that alleviates forgetting following classical conditioning. A sizable reduction occurred in the forgetting normally observed after a 4-week interval if the subjects (rats) had been reexposed, prior to testing, to the UCS of conditioning. For animals conditioned as young adults (about 80 days of age) or as weanlings (22–26 days of age), forgetting was alleviated whether the UCS was presented on several occasions throughout the retention interval (reinstatement paradigm) or on a single occasion 24 hours prior to retention test (reactivation paradigm). Neither treatment affected the forgetting of rats conditioned as infants (16 days of age), if the retention interval was 4 weeks in length; but after a shorter interval (7 days), forgetting was alleviated in these animals as well, with both paradigms.

To test whether this reactivation treatment enhanced retention scores through means unrelated to memory retrieval, by inducing significant new learning or by enhancing general activity in the animal, for example, several control experiments were conducted. These experiments yielded no indication of such a relatively trivial basis for the reactivation effects. The magnitude of the present reactivation effects suggested, moreover, that if a single factor of this trivial sort were responsible, it would be a fairly obvious one and its identity readily determined. Having failed to identify any such factor, we tentatively conclude that re-presenting the UCS alleviates forgetting by facilitating memory retrieval. We emphasize the tentative nature of this conclusion: studies of reactivation effects must continue to include, and even to expand, the full set of control conditions needed to isolate simple performance effects or consequences of new learning induced by reactivation treatments, if we are to avoid a ridiculous application of the retrieval construct to explain all bumps in a retention function.

In what sense may the reactivation effects studied here result from facilitation in the retrieval of memory attributes representing the episode of original conditioning? We have suggested that such facilitation may occur if the organism is provided access to stimuli that are represented as attributes of the critical memory but that otherwise are absent or not noticed by the organism. In terms of concrete operations, we have suggested that retrieval may be facilitated either by providing appropriate stimuli within the contemporary context of the retention test or by "prior cueing," presenting appropriate stimuli to the subject prior to the retention test. Although ample evidence for both instances may be cited (for reviews, see Spear, 1971, 1973, 1976), it is largely indirect and cannot be claimed as conclusive, especially in terms of the "pure" action of prior cueing. For example, prior cueing may be exemplified in the present use of the reactivation paradigm, but perhaps not exclusively. The effect of presenting the UCS 24 hr before the retention test conceivably may occur primarily through modification of the contemporary context of the test rather than through prior cueing alone. Possibly, presentation of the UCS initiates in the animal a long-lasting emotional change, which persists for 24 hr. If so, however, the long-lasting emotional change is unique to animals actually conditioned previously (as shown in the present experiments). This renders such an effect unlikely, unless it is mediated by the consequences of conditioning, which implies memory retrieval.

Changes in the contemporary context of testing may facilitate memory retrieval in terms of the consequential arousal of corresponding attributes of a sufficient number or kind to cause retrieval of either the entire memory (that is, redintegration, which seems a relatively unlikely mechanism), or of the specific response attribute needed to manifest the memory in behavior (which, within the present classical conditioning paradigm, can only be internal, for example, adrenal–pituitary response), or of the simultaneous arousal of the CS and UCS attributes. The mechanism for the influence of prior cueing must be different. We suggest that prior cueing results in the direct arousal (retrieval) of the attribute being cued (in this case, the UCS), perhaps together with associated attributes, leaving them in a state of accessibility similar to that found after original learning. With negligible sources of forgetting present between the reactivation treatment and the retention test, we then would expect performance on the retention test to be similar to that found shortly after original learning (for further discussion of these mechanisms, see Spear, 1976).

Ontogenetic influences on the effectiveness of a reactivation treatment are suggested by the data we have presented. As with any subject variable, however, conclusions concerning the effect of age per se may be accomplished only with great difficulty, if at all. We already have noted the problem of correlated differences in degree of learning, which is solvable. A more difficult problem concerns the absolute probability of memory retrieval following introduction of a source of forgetting. Immature animals seem more susceptible to certain sources of forgetting, such as long retention intervals (e.g., Campbell & Spear, 1972) and perhaps interference (e.g., Spear, Gordon, & Chiszar, 1972). If we find that reactivation treatments have less influence on the retention of younger animals, perhaps it is simply because they have forgotten more and that generally, the more effective a source of forgetting, the less effective a reactivation treatment. There is insufficient evidence to determine whether this is generally true however, and certain results show the opposite relationship (that is, greater benefit from aids to retrieval the greater the forgetting; Woods & Piercy, 1974).

It should be obvious that the study of alleviated forgetting is in its preliminary stages and that a good deal of empirical analysis is required for the variety of remaining questions. An obvious example concerns the action of the UCS as a reactivation treatment. Are the mechanisms underlying its influence on the alleviation of forgetting the same as those involved when re-presentation of the UCS alleviates the effects of extinction (Hoffman, 1969; Wilson, 1971; Rescorla & Heth, 1975)? Is the reactivation consequence of the UCS equal in instrumental and classical conditioning? Preliminary data from our laboratory indicate that the UCS, although a generally important attribute, may have relatively greater significance when an attribute representing an instrumental response has not been included (encoded) in the memory. Is the primary reactivation consequence of the UCS internal (neurohormonal) in nature, so that any severe stress with common internal consequences may substitute for the footshock UCS as a reactivation treatment? Preliminary data from our laboratory indicate that this may be the case, but the boundary conditions are unclear and the implications profound. For example, one might then be led to expect enhanced retention for a memory involving an aversive reinforcer if, throughout or following a source of forgetting, the organism were constantly bombarded by various stressful events having similar internal consequences; by comparison, poorer retention would be expected if a relatively tranquil life were led. Such a prediction must be modulated by consideration of whatever interference with retention may arise extraexperimentally from acquiring memories involving similar stress (cf. the "interference paradox," Underwood & Schulz, 1960). However, this borders on undisciplined speculation, a further indication of the preliminary nature of this research and the need for further systematic investigation.

ACKNOWLEDGMENTS

Preparation of this paper, and the research described herein, was supported by grants from the National Science Foundation (BMS 74-24194), the National Institute of Mental Health (MH 18619 and Research Scientist Development Award MH 47359), and the Rutgers University Research Council. The technical assistance of Norman G. Richter is gratefully acknowledged, as is the editorial advice of Charlotte M. Pastore, the statistical assistance of Harrel Wright, and the secretarial skills of Muriel Cohn. This paper has benefited from comments and suggestions provided by Robert C. Bolles, Alvin Berk, and Ralph R. Miller.

7

Memory and Coding Processes in Discrimination Learning

Roger T. Davis
Sally S. Fitts

Washington State University

INTRODUCTION

We had been studying cognitive processes in monkeys for some time (Davis, 1974; Medin & Davis, 1974) when we turned to the problem of memory for stimuli and their reward signs in discrimination learning paradigms. Generally the relationship between discrimination learning and memory paradigms has not been worked out well. For example, it has been common to think of discrimination learning essentially as a recognition test. If this were so, one might expect recognition memory in monkeys to show a capacity resembling that described by Standing, Conezio, and Haber (1970) for human beings. However, discrimination learning involves more than recognition. If each object in a series were associated with reward or nonreward in a nonsystematic manner the monkey likely could remember if the object had occurred before. In addition, however, when the stimulus is paired with another object, the animal must remember whether or not the original stimulus has been associated with reward in order to perform appropriately.

A simple self-administered example illustrates the difference between recognition capacity and the problem of remembering both the stimulus and its incentive value. Rapidly thumb through a well-illustrated magazine, looking at pictures but not reading the captions. Later, open the magazine to any page and ask yourself whether or not you previously have seen the picture on that page. An alternative way of conducting this demonstration is to note not only the picture, but also if it is on an odd or even numbered page. Later look at a picture and try to remember whether or not you have seen it before, and if so, if it was on an odd or even numbered page. The first demonstration can indicate a capacity of several hundred, if not several thousand, alternatives. The second becomes difficult for 20 pictures.

Accepting that the recall of discriminations is limited to a relatively small set of stimuli, we have asked whether or not discrimination performance is a direct consequence of list length, elapsed time, and incentive conditions. The following specific questions determined the course of the research:

1. Is there a functional relationship between recall and the number of successively presented stimuli?
2. If performance decays as the number of stimuli increases, is the decay caused by interference or by elapsed time?
3. Because discriminations involve reward as well as nonreward, are forgetting functions for the two incentive conditions basically similar or fundamentally different?

Parametric information on these questions should be a good beginning point for analysis of memory processes in discrimination learning.

REVIEW OF RELATED PARADIGMS

The procedure we used involved a combination of the traditional paradigms of concurrent discrimination, matching to sample, and delayed response [see reviews of the first two paradigms by French (1965) and the last by Fletcher (1965)]. Each of these paradigms, as well as object-quality discrimination learning, has been employed for some time to study memory for discriminations, but each one has disadvantages for answering the whole set of questions. We ultimately combined features of all three.

Concurrent discriminations involve successively presenting pairs of stimuli, with one member of the pair rewarded and the other not. When the set is exhausted, it is repeated with a differently randomized order of pairs until the subjects are able to make correct discriminations of every pair in the set on one or two successive repetitions. The procedure is directly analogous to verbal discrimination learning in human studies. Although concurrent discrimination problems have the advantage of presenting a number of stimuli before a test for memory, they do have several drawbacks. First of all the use of a criterion allows for overtraining of some stimuli. Second, discrimination consists of two simultaneously presented stimuli, each having a different reward sign. Therefore, it is difficult to assess the contribution of either reward or nonreward to the memory for any one object. Third, and following from the two previous drawbacks, unchosen nonrewarded objects accompanying rewarded objects gain in attractiveness as the number of pairings of the two stimuli increase (Leary, 1956; Fitzgerald & Davis, 1960). Finally, as Leary (1958a) recognized earlier, concurrent discrimination problems necessitate longer time intervals between discrimination trials than one may like.

The delayed matching to sample problem has features that, if incorporated into the concurrent paradigm, may solve some of the aforementioned problems.

Delayed matching usually requires a choice after a time delay between two or more stimuli, one of which previously has been presented singly and rewarded. This paradigm differs only slightly from the first two trials of discrimination problems in which a single stimulus is presented on the first trial in the manner described by Harlow and Hicks (1957). This kind of discrimination problem employs either reward or nonreward on the first (or sample) trial, unlike matching, which typically involves only a rewarded sample. The latter procedure implicitly assumes that memory functions for reward and nonreward are the same and, as we shall see, this is not a safe assumption.

Leary (1958b) studied the effects of reward and nonreward over time delays in a concurrent discrimination paradigm. His earlier work (Leary, 1956, 1958a) had convinced him that there were temporal changes in object attractiveness as a function of the conditions of reward. He (1958b) pursued this idea in a study of monkeys' performance on immediate (2-min) and delayed (24-hr) retention tests for stimuli presented one at a time and either rewarded, not rewarded, or alternatively rewarded and not rewarded on eight training repetitions of each of the lists of nine objects. The tests provided a discrimination between stimuli previously encountered in the list and new stimuli.

Leary's results showed far more errors in the immediate test than in the delayed test for the alternately rewarded stimuli, but about four times as many errors on the delayed as on the immediate test for nonrewarded stimuli. In the immediate test, nonrewarded stimuli were avoided more successfully than rewarded stimuli were chosen, but this demonstration of the Moss–Harlow effect was not found after 24 hr. This result contradicted an assumption by Harlow and Hicks (1957) that the delay functions of reward and nonreward were the same and provided impetus to examine these effects further in the present series of experiments.

Although Leary had shown that the distributions of the effects of reward and nonreward were different over time, his study did not give conclusive information about the nature of memory processes involved in these distributions. This was because he repeated the lists of single stimuli a number of times, making the assumptions that the effects of reward sum in a simple manner and that there is no difference in retention of stimulus patterns and reward signs.

PRESENT EXPERIMENTS

Assuming that the effects of reward signs and number of stimulus patterns are thus complicated by the temporal distribution of the effects of reward, we have tried to sort these variables out in a training series that have involved presenting 1, 2, 4, and ultimately 16 stimuli, one at a time, with a random order of reward signs for successive stimuli. After each list was presented once, there was a memory test, sometimes after short added delays, on which one of the stimuli from the list was paired with an unfamiliar stimulus. If the reward sign of an object was positive in the list, this object was again rewarded on the test; if it was not rewarded when presented singly, the alternative unfamiliar stimulus was rewarded on the test. The animal then proceeded to another list of stimuli, none of which was ever repeated after the test.

The several experimental procedures are outlined in the following section. These procedures varied the number of stimulus samples and the pattern of reward signs, stimulus positions, and the delays between sample presentation and test for memory. The general procedures are outlined first, followed by specific procedures and results.

GENERAL EXPERIMENTAL PROCEDURES

Subjects

Fourteen monkeys were employed as subjects in these experiments. Five were very old rhesus (*Macaca mulatta*), three were the old progeny of the first group, and six were adult pigtailed monkeys (*Macaca nemestrina*). The very old animals averaged 2 years of age when purchased and had been first at the University of South Dakota, and later at the Washington State University primate laboratory, for a total of 22 years, 4 months. Their progeny were laboratory born and were aged from 17 years, 4 months to 19 years, 2 months. The use of these animals has been detailed for many other experiments (Davis, 1974). The pigtailed monkeys were jungle born and were approximately 8 years old. They each had 5 years of experience in the laboratory, were females like the old animals, and were trained to the same criterion of performance at the beginning of the present series of experiments. Because there were no reliable group differences in these experiments, the data of all subjects were pooled.

Apparatus

The apparatus used in this series of experiments was another of the many modifications of the Wisconsin General Test Apparatus (WGTA). It was specially designed to provide rather short interstimulus intervals (ISIs) and the possibility of presenting succeeding trials uninterrupted by lowering an opaque screen. We have not used an opaque screen because the screen indicates that the trial has been terminated and consequently produces a deficit unrelated to what we commonly regard as forgetting (Motiff, DeKock, & Davis, 1959).

A formboard was mounted on a 45° slant, to each side of a four-sided frame that was positioned on a base that rotated smoothly on a ball bearing assembly about a vertical axis above the table of the WGTA. Each formboard had two standard foodwells placed 30.5 cm apart, and the experimenter could position the rewards in the correct foodwells and cover the foodwells in each formboard with stimuli before initiating a sequence of trials. Stimuli were obtained by cutting colored pictures out of magazines and gluing them to 5.3 × 5.3 cm squares of masonite. An attempt was made to include as many colors and textures as possible, to avoid pictures of familiar objects, and to provide unique stimuli for each problem.

The mean time to make particular operations with the apparatus was determined, separately from the present procedures, for one-, two-, and four-sample problems. This enabled a number of internal comparisons to be made between various procedures. The time taken to rotate the device 90° from one formboard to the next averaged approximately 1.0 sec (between samples in four-sample problems and from the last sample to test in two-sample problems). It took 1.5 sec to rotate the formboards 180° on the one-sample problems. A 1.06 sec sample-to-test time for four-sample problems involved flipping a second board down from the top and was comparable in duration to rotating the apparatus 90°. For 16 samples there was approximately 5 sec between samples and about 1 min between problems.

Procedures

There were eight different experimental procedures, most conveniently grouped as three experiments. The first experiment was concerned with the effects of varying the number of samples and with possible serial position effects. The second one was a control for the first and determined the effects of presenting unfilled time intervals similar to those involved in the first. Experiment 3 looked at the effects of reward and nonreward on memory for discriminations. Because repetition of two of the simple procedures at various stages of the experimental program has failed to indicate any practice effects or reliable decline in performance level, we shall simplify our presentation here by not presenting a procedure by procedure chronicle.

The basic design of the procedures involved presenting a group of similar problems, every one of which consisted of one or more sample trials and a retention (matching) test. A sample trial consisted of presenting a stimulus picture over either an empty or a rewarded foodwell on the left- or right-hand side of a formboard. The remaining foodwell was empty and was not covered. A given picture was never used on more than one sample trial and was never repeated between problems, so there was only one opportunity for an animal to associate the sample with a reward or nonreward.

A retention test presented two stimuli, one that had occurred as a sample within the problem and one that was unique. If the sample had been rewarded, it was rewarded again if it appeared on the retention test; if it had not been rewarded as a sample, however, the unique stimulus rather than the previous sample was rewarded on the retention test. The variables that constituted the several experiments are detailed below.

SPECIFIC PROCEDURES AND RESULTS

We shall group together procedures that have a common purpose. However, some reference is made to the order of particular procedures to illustrate that animals soon reach an asymptotic level of performance and that data points are typically recovered through repeating procedures. What we refer to as the first experiment has been concerned with the effect of the number of samples on test performance. The second experiment was also distributed throughout the various procedures and was concerned with whether or not decay in performance was dependent on elapsed time comparable to that involved in presenting stimuli. The third experiment dealt with the effects of reward and nonreward and how these effects distributed over filled and unfilled intervals of time.

Experiment 1: Number of Samples

Does performance deteriorate as a function of the number of samples in the list, and, if so, is there a serial position effect? We investigated this question by first training the animals to remember a singly presented sample over a brief interval in the manner described in the general procedures. Every animal was given 6 days of baseline training and each met a criterion of 18 correct responses out of 24 on two consecutive days. They then proceeded to successive procedures in which the number of samples preceding the test was first two, then four. Animals were then returned to the condition of a single sample before the test but with four possible delays (0, 2, 5, or 20 sec) between the two trials of the problem. Subsequent procedures provided two then four samples. Finally there were two procedures that provided 16 samples before various tests, and these procedures were separated by one-sample problems to determine whether practice effects had occurred

Figure 7.1 shows that the performance of monkeys worsens as a function of the number of samples preceding a test. The upper curve with open circles represents the percentage of correct responses on procedures that did not include a delay between the last sample and the retention test. The lower curve was obtained on procedures with delays interposed between the last sample and the test. The solid line represents an average of the results for from one to four samples. Procedures having 16 samples did not involve the presentation of additional delays between the last sample and the test, although the durations between some of the samples and the test were necessarily long.

There was an obvious decline in performance as a function of the number of stimuli and the effect was sufficiently robust to occur both with and without added delays. However, the fact that forgetting occurs even with a single stimulus necessitates looking at both serial position retroactive inhibition (RI) effects and the effects of delays per se.

Figure 7.2 indicates that there are considerable serial position effects and that

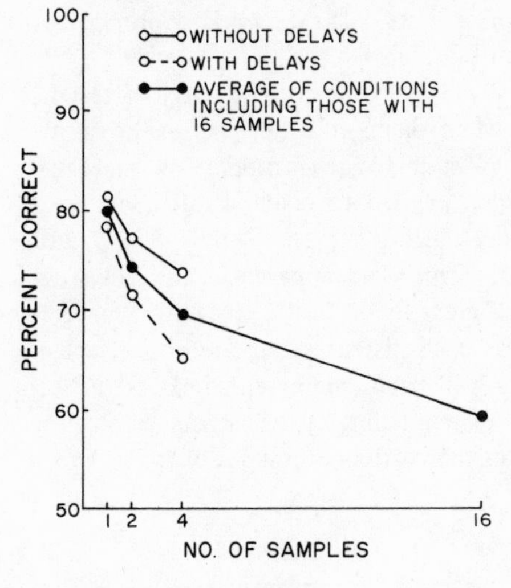

FIG. 7.1. Performance of monkeys on various numbers of samples as a function of the presence (○– – –○) or absence (○——○) of added delay conditions. Solid circles show average conditions, including those with 16 samples.

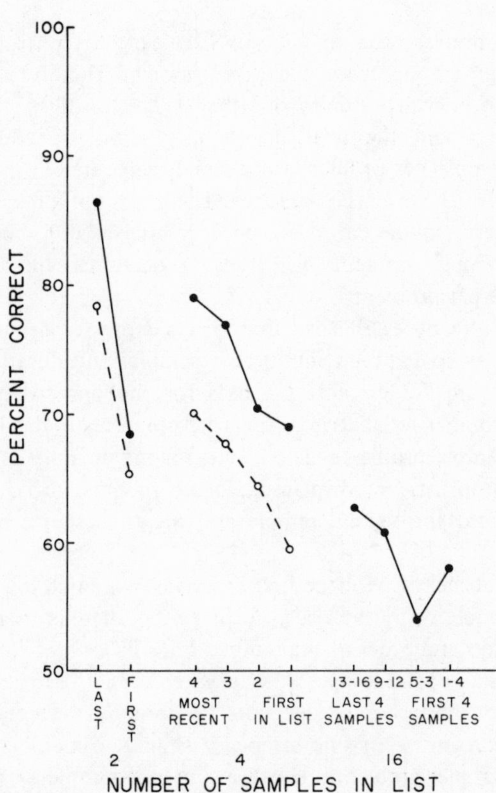

FIG. 7.2. Effect of ordinal position of samples and delays on memory by monkeys. (●) Without delays, (○) with delays.

these are most pronounced for the shortest lists. Also shown is poorer performance when delays have followed the last sample.

The serial position effects were statistically reliable using an ANOV test for two- and four-sample procedures, both when there were delays after the sample and when there were no delays. The F value for performance on successive blocks of four trials in 16-sample problems failed to reach significance, but a questionably admissable t test for recency, which compared Samples 13–16 with Samples 5–8 was significant at the .02 level. The apparent trend toward a primacy effect, which would have been of great theoretical interest, was far from statistical reliability. Although there was a consistent serial position effect, it diminished as the number of samples per problem increased. It is unclear whether this lessened effect was a result of weakening of the original association or merely of the time taken to administer the particular number of trials. This is clarified in the sections that follow.

Experiment 2: Effects of Delays

The previous section showed that it was necessary to control for time delay when we were presenting more than one sample. Therefore, we introduced several procedures in which a delay of either 0, 2, 5, or 20 sec separated the last sample in a series and the test for memory. Two procedures with delays employing one sample per problem were widely separated in the experiments in order to control for the effects of practice. There was not a reliable difference in the results of these, so the data were pooled. Interpolated between the procedures with one-sample problems were delay procedures with two samples and with four samples per problem.

Figure 7.2 had already established that delays depressed performance, but the functional relationship between number of samples and duration of delay was not indicated. Figure 7.3 presents the data for one- and two-sample problems and includes a comparison between the performance on the first and second samples of two-sample problems. The results for the memory of the first sample of two-sample problems are displaced 2.5 sec to take into account the time necessary to present the second sample and the time that a monkey takes to respond.

It is obvious that memory for the first sample is at a much lower level than the second at comparable delay intervals. Furthermore, there is a significant interaction between delays and order of the sample, $F(3, 27) = 5.667, p < .01$, because performance for the second sample is better after 2- and 5-sec delays than after 0 or 20 sec, and the opposite trend occurs for samples occurring first in order. The function for memory of one sample is similar to memory for the second sample of two-sample problems. However, in one-sample problems only the difference between 5- and 20-sec delays is statistically reliable. We wondered whether this reflected a change in the probability of going to the previously

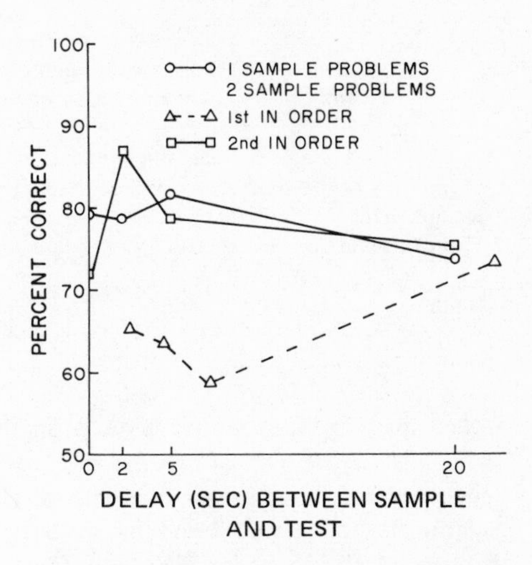

FIG. 7.3. Retention of samples in one- and two-sample problems as a function of the delay between the last sample and the test. There is an additional 2.5-sec lag for the first sample of the two-sample problem (1.5 sec operating time; 1.0 sec response time) because of the presence of the second sample.

rewarded object, position, or some combination thereof. It did not, although there was a statistically reliable increase in responses to the previously nonrewarded object in the previously nonrewarded position from 8.3 to 12.3% at 5 and 20 sec, respectively. This does not exhaust all of the possibilities for forgetting and only accounts for some of the decline in performance.

These data suggest a large RI effect, which dissipates. However, merely saying that there is RI does not identify its source. The interference could be some sort of simple stimulus confusion, delay of consolidation, or maybe an interaction between order of stimulus presentation and particular incentive variables. We chose to look at the last possibility in the third experiment.

Experiment 3: Rewards and Memory

To examine the effect of incentive and how it interacted with delay and stimulus order we compared conditions in which two-sample problems were given both with and without delays after the second sample. The order of the sample should be the only variable related to memory if there is interference between the characteristics of the stimuli but not incentives. If interference is a function of confusions in remembering reward signs, however, then performance should be worse if the reward signs of the two samples are dissimilar than if they are the same.

Table 7.1 indicates that the pattern of results in conditions with and without delays is quite similar in that performance is much better if the reward signs are alike rather than different. The interaction between sign of the first sample and sign of the second sample is significant both with and without delay, $F(1, 9) = 179.276$, $p < .001$; $F(1, 9) = 16.687$, $p < .01$, respectively. Also, as we have noted in the previous section, performance is considerably better if the second

TABLE 7.1
Retention of the First and Second Samples in Procedures with and without Delays as a Function of the Reward Sign of the Sample

	First sample retention				Second sample retention			
Sign of first	+	–	+	–	+	–	+	–
Sign of second	+	–	–	+	+	–	–	+
No delays	82.00	82.67	48.67	60.00	86.53	89.11	84.77	84.45
With delays	72.50	82.50	51.88	55.00	76.25	80.00	80.00	76.25

rather than the first sample is used on the retention test. This effect is very striking when the signs of the first and second samples are dissimilar, but it is not remarkable, although characteristic of every monkey, for unchanging signs. Significant effects were found both with and without delays for (a) sample order, $F(1, 9) = 81.225$, $p < .001$, and $F(1, 9) = 29.434$, $p < .001$, respectively; (b) the interaction between sign of the first sample and sign of the second sample, $F(1, 9) = 179.226, p < .001$, and $F(1, 9) = 16.687, p < .01$, respectively; and (c) the second-order interaction between order of the sample, reward of the first sample, and reward of the second sample, $F(1, 9) = 114.572, p < .001$, and $F(1, 9) = 13.751, p < .01$, respectively. The last named interaction is the data of Table 7.1, and it shows that the patterns of reward signs and sample order dramatically interact to determine what is remembered. No other terms were significant in the procedure without delays, but sign of the first sample was significant at the .05 level in the procedure with delays. Significant interaction effects of delay with reward sign also occurred and are noted in the following paragraphs.

The reliable difference in performance between conditions under which both of the samples have the same reward sign and conditions under which they are different decreases with delay and, like the effect of sample order, is virtually absent after a 20-sec retention interval. Apparently, whatever is increasing or decreasing performance has ceased to be effective, so it is hard to imagine that there is much interference between samples. Also, one expects little problem-to-problem interference using these procedures because the interval between problems is about as long as the delay periods.

Figure 7.4 offers another tantalizing effect. Whereas performance on the retention trial following two rewarded samples is nearly the same as that after two nonrewarded samples for short delays, it is considerably better after two nonrewarded samples than after two rewarded samples for longer delays of 5 and 20 sec. This is, of course, in the range of successive trials in the WGTA and suggests that the Moss–Harlow effect is related to time-dependent effects of reward and nonreward. As we indicated before, the results with one-sample problems are not as complex, undoubtedly reflecting the importance of succeeding samples and signs to this effect. Animals perform better on problems with

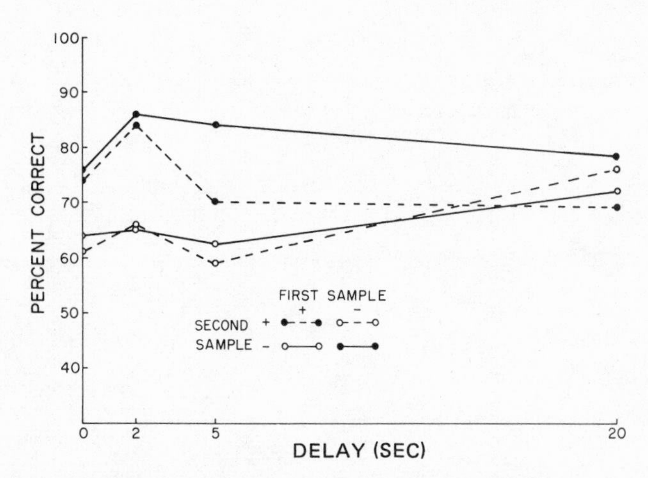

FIG. 7.4 Signs of the first and second samples in a two-sample problem as a function of delays between the last sample and the test. In the key a plus sign designates reward and minus represents nonreward.

one nonrewarded than on ones with one rewarded sample (81.6 and 75.2% correct, respectively), and the effect is relatively constant between a 5.0 and 6.8% difference for the three shorter delays. However at 20-sec delays, when performance drops for one-sample problems, as was shown in Fig. 7.3, the decline for the rewarded sample is 13.8% greater than that for the rewarded sample.

Therefore, pursuing the interactive effects with two-sample problems seemed more promising. This is borne out in Fig. 7.5 which represents the combination of the effects in Figs. 7.3 and 7.4. Rather dramatically, it shows that the performance changes are related to both the incentive conditions and the sample order. Performance reaches 100% at 2-sec postsample delay with both of the samples rewarded. This function is more striking than the previously noted distribution of performance over time in procedures with one sample per problem.

Maybe even more remarkable than this momentarily efficient performance is the performance following unlike signs on samples when the first of the samples is tested. It begins at chance, but after 20 sec attains a level over 67% correct. One might explain these results as rehearsal of the object with the changed sign and interference with the one preceding it, assuming that the simple decline in performance for one-sample problems between 5 and 20 sec occurred because there was no second sample to rehearse. However, a simpler explanation should be found.

It is also possible that the results bear some resemblance to the complex findings by Robbins and Bush (1973) on chimpanzees' memory for discriminations with other ones interpolated. They reported a decline in Trial 2 perfor-

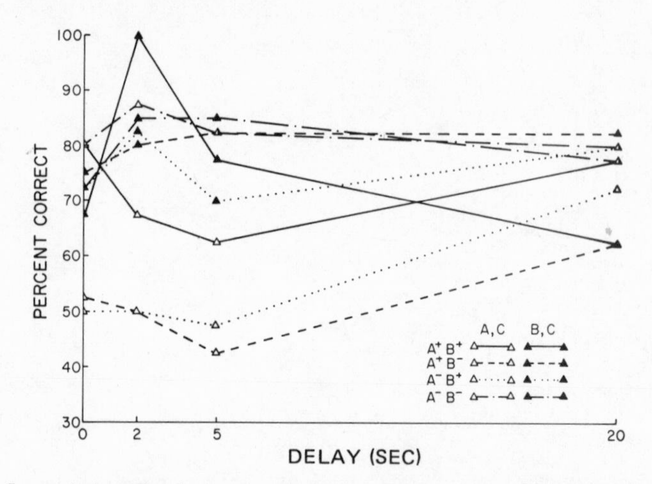

FIG. 7.5. Interrelationships between order of samples in a two-sample problem, the signs of the samples, and the delays between the last sample and the test. In the key, A designates the first sample, B represents the second, and C the new stimulus presented on the test with either A or B. The left-hand column indicates the reward sign of each sample, the next one indicates conditions of the first sample on the test, and the last column contains conditions on which the second sample occurs on a test.

mance as a function of the number of interpolated stimuli. Performance on Trial 3 decreased as a function of the number of interpolated stimuli between Trials 1 and 2, if only two stimuli were interpolated between Trials 2 and 3; however, if 10 stimuli were interpolated between Trials 2 and 3, performance on Trial 3 increased as a function of the number of interpolated stimuli between Trials 1 and 2. However, these results are not closely related to our own because we do not know the contributions to performance of reward sign, temporal intervals, sample size of available stimuli, or possible estimates that the animals may have made of time intervals in the Robbins and Bush (1973) study.

DISCUSSION

These experiments were designed to provide information about memory for singly presented stimuli rather than to be a theoretical *tour de force.* In this section we relate the main findings to the questions we have posed earlier in this chapter and to the results of similar experiments.

Performance decreased as the number of samples increased in a list, corroborating our notion that something more than simple, nearly infinite recognition was involved in this task. This conclusion led us to several other questions and some very puzzling results concerning the nature of interference and the interactions between number of stimuli, time delays, and incentive conditions.

Briefly, we found the following:

a. Performance decreased as the number of samples before a test increased.

b. Within a given sample set size, performance showed a pronounced recency effect, favoring the latter samples in a sequence.

c. This effect was specially large if reward signs changed between the first and second of two samples.

d. There was a complex relationship between the duration of the delays and the order of the samples; performance was better at 2- and 5-sec delays than at 0- or 20-sec delays for the most recently presented samples and in the opposite direction for the preceding samples.

e. Sample sign also interacted with the delay interval. Performance was poor at short intervals but got better at longer intervals for those problems with changing signs between the first and second sample.

Performance was uniformly good for problems where the signs were the same on both samples. Understandably, the effects in d and e were synergic, and there was a dramatic range from 50 to 100% correct depending on the four variables: delay, order of the sample, sign of the first sample, and sign of the second sample (refer to Fig. 7.5).

There are correspondences between our results and previous work on memory and discrimination learning in monkeys. For example, short-term retention of discriminations has been studied by varying the intertrial interval. The idea of such experiments is that the initial choice can be forgotten before the second and subsequent trials are presented. The decline is modest for 10–60-sec delays with overall performance varying around the 90% level (Riopelle & Churukian, 1958; Harlow & Warren, 1952). In the present studies tests for single samples or for the second of two samples is in the range of 80–85% correct, and, if neither the first nor the second sample is rewarded, the level of performance declines very gradually from 2 to 20 sec (see Fig. 7.4). A similar comparison can be made to the declining memory functions in delayed matching to sample during the presence of house light (D'Amato, 1973).

However, other of our results are surprising and difficult to understand. Consider the results on retroactive interferences as a function of retention interval (Fig. 7.3). Can interference theory account for this pattern of results? Classically McGeoch (1942) held that RI involved incompatable responses being associated with the same stimulus. Suppose for a moment we take a traditional view that reward leads to an approach response and nonreward leads to a competing response. In our two-stimulus problem, performance is the poorest when the two samples are differently rewarded, which may produce response competition. However, performance on the first and second samples of a pair does not show the reciprocal function one expects because there is little proactive interference. Neither could a simple interference theory handle the first increasing, then decreasing interference as the delay interval increased.

Even an elaborated interference theory adding spontaneous recovery and unlearning to its conceptual armaments does not seem able to untangle the pattern of results shown in Fig. 7.3.

Equally puzzling are several of the simple delay functions. There is an anamolous peak in the retention of the most recently presented sample at 2 sec for two-sample and later for one-sample problems. This effect is difficult to relate to others, but it is interesting that the higher values obtained are similar to those obtained in memory for discriminations and in delayed matching (e.g., Robbins & Bush, 1973). At this moment we do not even understand the "simple" delay functions.

Finally, we consider nonreward on memory. This was not remarkable when only one sample was employed, but when two samples occurred in a problem the effects were very striking. The effects of reward and nonreward in sequences change completely as the delay interval increases and other studies showing mixed results may have differed in the delay intervals involved. For example, as we mentioned earlier, the Moss–Harlow (1947) effect could very well be a result of the particular interval between presenting the stimulus singly and the first discrimination trial. Although Harlow and Hicks (1957) assume that the distribution of reward effects over time delays resembles that of nonreward effects, both Leary's (1958b) results with memory for concurrent discriminations and the present studies indicate that this occurs only at certain poststimulus time intervals.

Our parametric investigations have not answered many questions concerning memory processes in discrimination learning but they caution against hasty generalizations. The considerable novelty of certain of our results removes much of the staleness from the usual call for more research. We obviously need much more research on memory processes in discrimination learning.

ACKNOWLEDGMENTS

This research was conducted at the Primate Laboratory of Washington State University and was supported in part by United States Public Health Service Grant HD05902-04.

8

Storage and Utilization of Information within a Discrimination Trial

Raymond T. Bartus

Psychopharmacology
Parke-Davis Research Labs

T. E. LeVere

North Carolina State University

INTRODUCTION

Most behavior involves differential responding. Because differential responding is often the result of differential environmental conditions, an appreciation for how an individual distinguishes, assimilates, and reacts to different environmental events is prerequisite for an understanding of behavior. This chapter therefore develops some ideas regarding the processes that may be involved in discrimination learning where differential information from the environment must be used to successfully execute alternative responses.

Of particular concern is the viability of some rather time-tested assumptions concerning information encoding and utilization during discrimination learning. For example, does the stimulus information that is temporally closest to a response necessarily become associated with that response, as classic S–R contiguity theory maintains? Also, is choice behavior exclusively controlled by the stimulus information that has the greatest history of past reinforcement, or do other variables within the discrimination trial exert an even greater influence? Finally, do rewards and nonrewards have functions other than strengthening or weakening associations? At the heart of these questions is the broader issue of whether the information-processing activities of animals should be viewed as basically inflexible and amenable to treatment in simple, mechanistic terms, or whether these processes should be seen as much less rigid and capable of being modified to suit the demands of particular experimental conditions.

SOME METHODOLOGY AND A FEW DEFINITIONS

The present report is mainly concerned with the acquisition of two-choice visual discrimination problems in rhesus monkeys (*Macaca mulatta*). We are concerned in particular with information encoding and/or utilization within the discrimination trial. A discrimination trial is defined on the basis of the primate's behavior and is represented by the period of time the monkey spends "looking at" the discriminative cues, when this looking culminates in an overt choice response. Looking that does not end with a choice response is not considered a discrimination trial because, in the simultaneous two-choice discrimination situation, the absence of a choice response indicates that the discriminative cues have not been controlling behavior.

Within a discrimination trial, we define three periods: (1) the prechoice response period, (2) the choice response period, and (3) the postchoice response period. The pre- and postresponse periods are of principal interest in determining how a primate uses the information provided by the discriminative cues. To set the stage for our later discussion, we define the prechoice response period as that

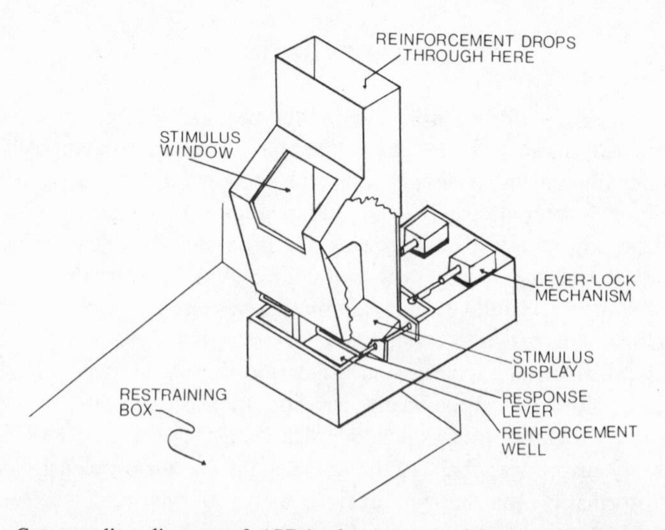

FIG. 8.1 Cutaway line diagram of APDA, the Automated Primate Discrimination Apparatus, used to collect the majority of data discussed in this chapter. In this version of APDA, the monkey sits in a plastic restraining cage mounted to the front of the apparatus. A pair of feeders (not shown), mounted directly above the apparatus, drop reinforcement pellets onto one of the two stimulus displays when a correct response is made. The reinforcement then rolls into the reinforcement well associated with the response level pulled.

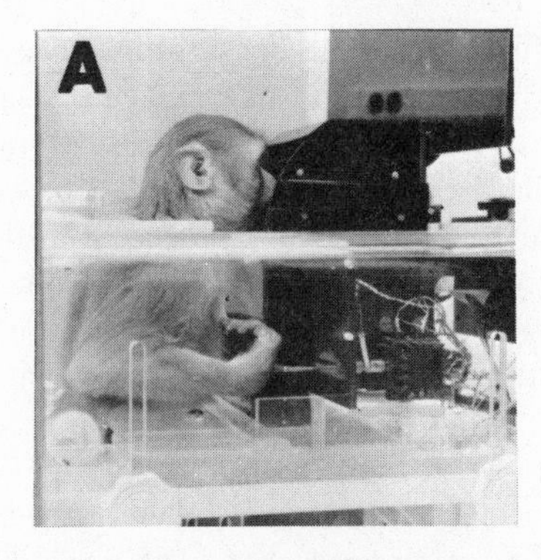

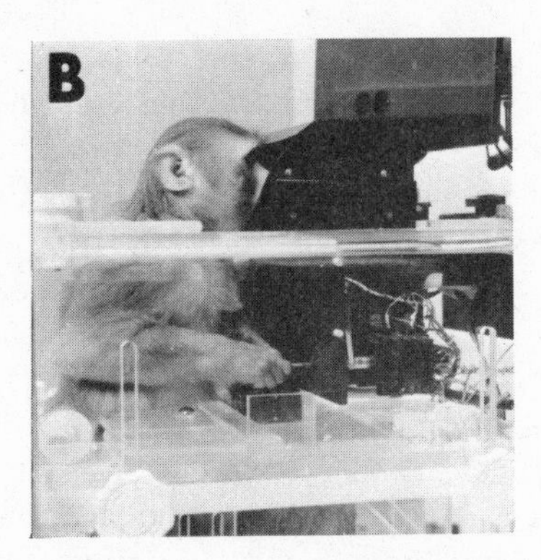

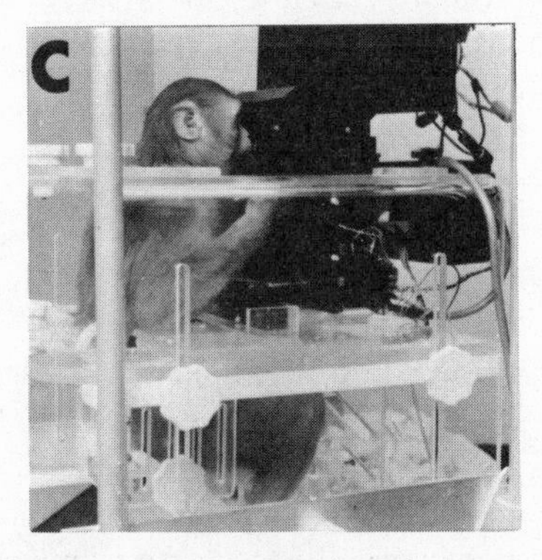

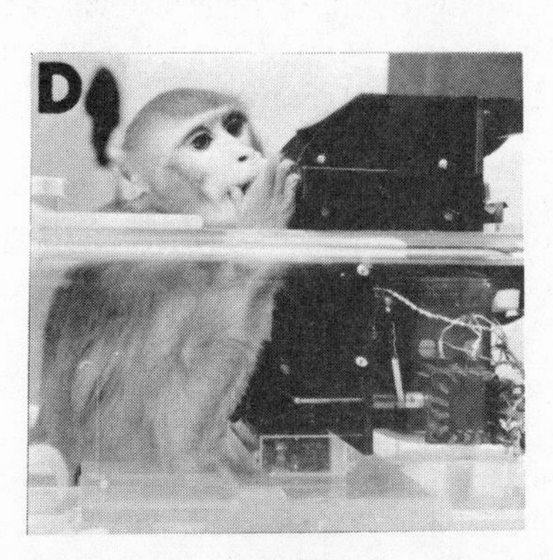

FIG. 8.2 Pictorial chronogram of monkey in current version of APDA. In this version, APDA is mounted in a Plas-Labs Primate Restraining Chair, and the monkey is seated directly in front of the stimulus observation window. The sequential intratrial events are as follows: (a) monkey observing stimuli through stimulus window during preresponse period; (b) monkey executing choice response by pulling right-hand response lever; (c) monkey continuing to observe stimuli during postresponse period; and (d) monkey placing reinforcement in mouth following correct choice response.

portion of the discrimination trial devoted to the determination of a particular choice response and involving (1) stimulus selection and (2) utilization of the stimulus information selected. Stimulus selection is simply defined as those events responsible for determining what information is eventually processed. Utilization of this information involves all those stimulus-related processes that determine which overt choice response is finally made. In the postresponse period, subjects are assumed to evaluate the effectiveness of their choice strategy, based on the outcome of the trial. Behavior during this period can, and often does, involve further interaction with available extrinsic stimulus information.

The data reported here have been collected using the Automated Primate Discrimination Apparatus (Figs. 8.1 and 8.2; also, see LeVere and Bartus, 1969). One salient feature of this apparatus is a stimulus observation window that permits more exact control of the stimulus conditions. For example, when patterns of electric light bulbs are used as stimuli, the monkey is required to place his face into the window and look toward the stimuli before the lights are turned on and the trial is initiated. Unless he maintains this "looking" behavior throughout the duration of the trial, including the time the overt choice is made, the trial is immediately terminated. Also, one can force the monkey to continue looking at the stimuli for any definable period of time by locking the response levers via electromechanical solenoids. Finally, the monkey's hands are visually isolated from the stimuli while his face is in the window, so that appropriate stimulus sampling, without obstruction and distraction of the monkey's own hands, is assured. As is to be seen, these features of the apparatus allow certain manipulations of stimulus information to be performed that have not been possible with more conventional testing procedures, such as the WGTA. We believe data obtained from these manipulations suggest the possible existence of certain intratrial mechanisms which heretofore have not been recognized.

A WORD ON STIMULUS SELECTION AND STIMULUS UTILIZATION

Ample evidence exists that stimulus selection within a discrimination closely corresponds to where the monkey looks prior to a choice response. Traditionally, it has been accepted that there are three independent determinants of where monkeys look prior to a choice response in the simultaneous, two-choice, visual discrimination paradigm: stimulus–response spatial contiguity, stimulus–reinforcement spatial contiguity, and cue size. The influence of these variables on stimulus sampling and the resultant effect on discrimination performance have been analyzed by Meyer, Treichler, and Meyer (1965) and by Stollnitz (1965). Although new data have been added (Schrier, 1970; Oscar-Berman, Heywood, & Gross, 1971; Schrier & Wing, 1973; Vaughan & Schrier, 1970; LeVere, 1968a, b) the message of these early analyses still appears valid. (A notable exception, however, may be the independent effects of cue size; LeVere, 1968b).

Other parameters may also exert an influence on stimulus selection. For example, the individual's prior experience, as well as stimulus complexity, cue dominance, and the like, all appear to be potential determinants. Although these variables may certainly be important in the final analysis, they do not appear directly related to those questions we wish to consider. For this reason they are not given further discussion in this chapter.

Once the monkey has isolated certain aspects of the stimulus situation, we may consider how this information is used to generate a choice response and how the information is modified by trial outcomes. One may suggest that the primate responds to the discriminative cue it has selected and that acquisition progresses, to some degree, in accordance with the law of effect. That is, reinforced choice responses are strengthened and nonreinforced choice responses are weakened. In either case, the discrimination trial and the monkey's interaction with the available stimulus information are assumed to end following a choice response and the occurrence or nonoccurrence of reinforcement. However, as we shall see, the acquisition of a discrimination is not readily amenable to any such tidy description. Rather, the monkey does not simply respond to one of the discriminative cues, nor does his interaction with the available stimulus information end with the occurrence or nonoccurrence of reinforcement. We first consider the prechoice response period, and specifically discuss some of the variables involved with stimulus utilization prior to a choice response. The question of whether or not information processing ends with the occurrence of a reinforcement is discussed in the following section.

PRECHOICE RESPONSE PROCESSES

Three of the questions we have previously raised are of particular importance when the monkey's prechoice response utilization of stimulus information is investigated. First, are the stimulus events (information) closest in time to the outcome of a particular response most likely to become associated with that outcome? Second, is the individual's choice behavior inflexibly controlled by the stimulus information having the greatest history of reinforcement? Third, can a primate modify how it uses the available stimulus information to meet the demands of discriminative situations? With respect to the first question, the dictates of classical association theory and data from eye movement studies (Oscar-Berman *et al.,* 1971; Schrier & Wing, 1973) seem to imply an affirmative answer—events temporally closest to an outcome are associated with the outcome. However, the above research does not provide a conclusive answer because the data have always been collected in situations where the stimulus information has remained static or unchanged prior to the choice response. By definition, the last visual stimulus that the monkey looks at necessarily provides the information that controls its behavior if this is the only discriminative information available prior to the choice response. If, in contrast, the monkey is sequentially exposed to more than one kind of visual stimuli prior to a choice response, the matter may be quite different.

An early indication that dynamic prechoice response stimulus information may produce behavior contrary to classical association theory and eye movement data stems from research by Polidora and Fletcher (1964). This investigation evaluated stimulus–response spatial contiguity in an automated version of the Wisconsin General Test Apparatus. Two groups of monkeys were trained; one group indicated their choice responses by directly pressing one of the two vertically oriented stimulus displays, and the other group indicated their choices by pressing one of two remote manipulanda below the stimulus displays. The results indicated that the monkeys which responded directly to the stimulus displays were significantly more proficient in solving the simultaneous two-choice pattern problem. However, in a second phase of the experiment, when the two groups of monkeys were required to exchange their modes of responding, a considerably different set of results occurred. The effect of modifying the response modes for the two groups depended on whether the individual monkey complied with the change in response rules as set down by Polidora and Fletcher. For example, if a monkey originally trained to respond directly to the stimulus display now responded to the remote manipulandum (as intended by the experimenters), its discrimination proficiency suffered. Similarly, monkeys originally trained to the remote manipulandum showed improved learning rates if they shifted to the stimulus displays during this phase of training (as intended by the experimenters). However, contrary to the experimenter's desires, some monkeys perseverated their original response modes and these animals showed little change in their discriminatory performance. If a monkey was originally trained to respond to the stimulus displays and continued to do so before it made the required response to the remote manipulandum, its discriminatory performance showed little decrement. If, in contrast, a monkey originally trained to respond to the remote manipulandum continued to attempt this response before it responded directly to the stimulus display, its discriminatory performance showed little improvement. In the case of these "double-responding monkeys," therefore, the first information the subject assimilated appeared to govern his choice. Subsequent information, even though temporally closer to the choice response, and in one case more relevant, appeared to be ignored.

We have more recently performed a more direct examination of the possibility that primates preferentially use the stimulus information occurring at the initiation of the discriminatory trial (LeVere & Bartus, 1971). This experiment used dot patterns as discriminative cues and trained Rhesus monkeys in the Automated Primate Discrimination Apparatus (APDA) discussed earlier. In this study, the control logic was programmed to require the observing response to be maintained for 1 sec prior to a choice response. This 1-sec prechoice response period was further programmed into two 500-msec subparts. During one of these 500-msec subparts, the subject was presented with a dot pattern relevant to the solution of the discrimination problem, whereas during the other 500-msec subpart the subject was presented with a dot pattern irrelevant to the solution of the discrimination problem. (Examples of these patterns and others discussed later are shown in Fig. 8.3.) The main question in this research was whether learning would be more proficient when relevant information occurred during the early 500-msec portion or during the late 500-msec portion of the required 1000-msec preresponse stimulus observation period. The results are summarized in Fig. 8.4, which depicts the mean number of trials to obtain successively more stringent criteria. Learning was most proficient when the subjects were trained on discrimination problems providing relevant information during the early

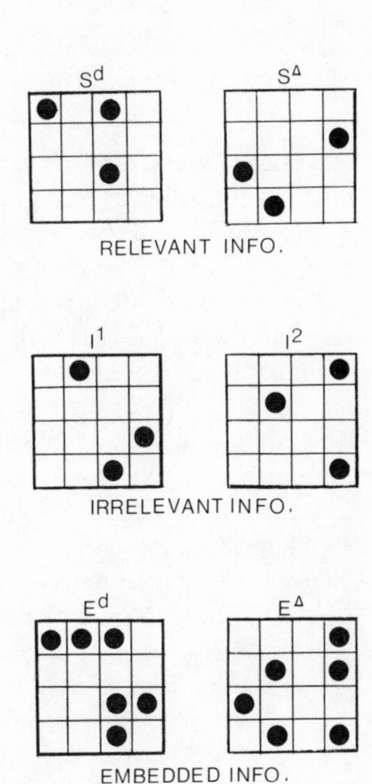

FIG. 8.3 Schematic representation of the type of stimulus information used in many of the studies discussed in this chapter. Each stimulus display consisted of a 4 × 4 matrix of .5-inch light bulbs. Discrimination patterns were formed by lighting three bulbs in each display (represented by darkened discs in figure). Examples of relevant information, irrelevant information, and embedded information (relevant and irrelevant superimposed) are shown in the figure and discussed in the text.

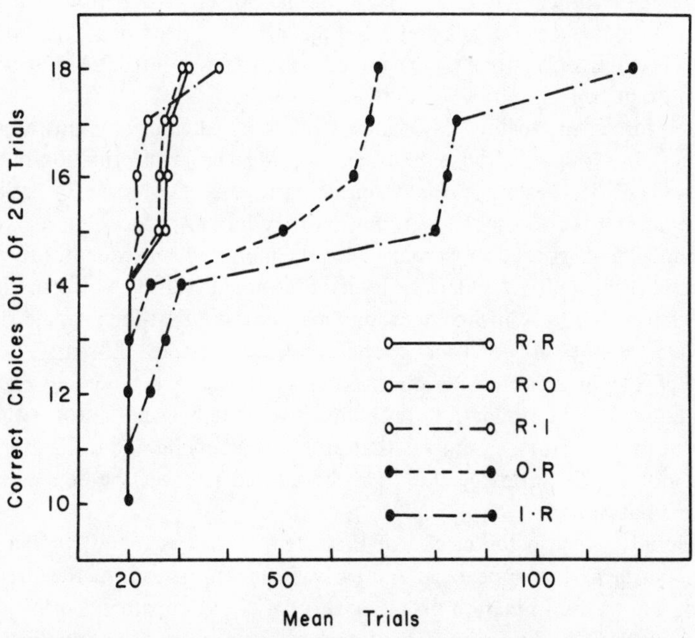

FIG. 8.4 The influence of dynamic stimulus information during the prechoice response period. The first letter of each curve label stands for the information presented during the first 500-msec portion of the prechoice response period and the second letter stands for the information presented during the second 500-msec portion. R, relevant information; 1, irrelevant information; and 0, no information (i.e., off).

500-msec subpart. Any discrimination problem that presented irrelevant information (dot patterns) or no information during the early 500-msec subpart was only mastered with some degree of difficulty. Perhaps most importantly, moreover, this difficulty was not attenuated when relevant information was provided during the later 500-msec subpart immediately preceding the opportunity to respond. Clearly, at least early in learning, the information closest to the choice response does not control the choice response. If the event closest to the reinforcement is not the sole determinant of what is learned, perhaps information utilization is more flexible than the traditional view of discrimination learning implies. We now turn to this question.

One form of asking this question is to examine whether the bias toward sampling and using early information occurs even when the monkey presumably "knows" which discriminative cues are relevant to the solution of the discrimination problem. LeVere and Bartus (1972) trained monkeys on a discrimination problem that presented only relevant discrimination information (dot patterns again) during both the early and late 500-msec portions of the prechoice response period. After the monkeys had mastered this discrimination, they were trained with this same relevant information available only in either the early or the late 500-msec portion. Either no information or irrelevant information was provided in the other 500-msec portion of the preresponse period. The results, shown in the right-hand portion of Fig. 8.5, indicated that the monkeys' performance fell only when irrelevant information was introduced in the early preresponse period.

The reinforcement history associated with the particular stimulus event does not necessarily override the subhuman primates basic information-processing strategy. That is, despite prior training with the relevant information, the monkeys attempted to use the information they initially processed, regardless of its relevance, and ignored or perhaps actively inhibited subsequent information. Because it is difficult to understand why the monkey ignores information that has previously led to reinforcement, one may propose that the primate simply is not aware of the information presented subsequent to the early 500-msec portion of the preresponse period. At the same time, the monkey eventually does master the discrimination problems with late relevant information and, therefore, the data further suggest that stimulus sampling and utilization strategies are not totally inflexible but may be altered to meet the demands of the learning situation.

One inconsistency in the results of these two studies is the effect that occurs when no information appears in the early 500-msec period. In the first study, involving a new discrimination problem, this condition impaired acquisition (Fig. 8.4; treatment O-R). In contrast, no detrimental effect on discriminatory performance was observed on this treatment in the second study, where the monkeys were first trained to criterion with the relevant cues (Fig. 8.5; treatment O-R). Either some subtle shift in processing strategy or perhaps a change in susceptibil-

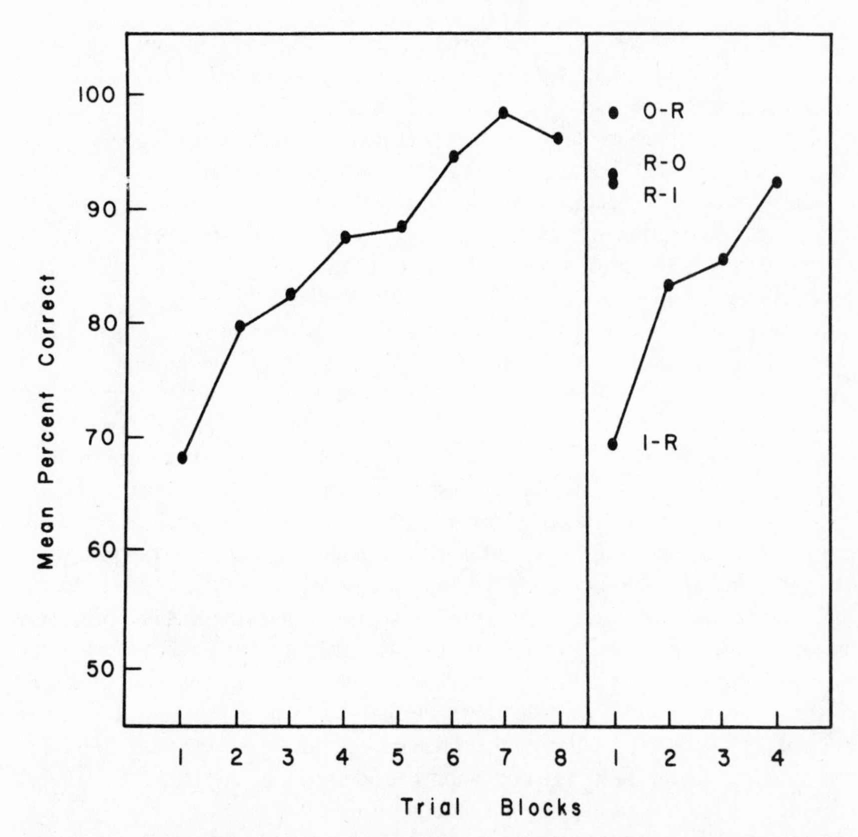

FIG. 8.5 The effects of introducing irrelevant information during the early or late 500-msec portion of the preresponse period (right panel) after the subjects had been trained with only the relevant information during the entire preresponse period (left panel). A trial block represents 20 correctional trials. The points and curve on the right panel are labeled as in Fig. 8.4.

ity to the effects of dynamic information may occur once the monkey knows what cue or cues are associated with reinforcement. Notwithstanding this subtle difference, however, by far the most dramatic effect of these studies is the consistent deficit found when irrelevant information is presented during the early segment, regardless of whether prior training with the relevant information has been given.

At first glance this powerful tendency to preferentially use the early information seems to argue against the more classical notions of association theory. However, if contiguity is defined subjectively rather than objectively, these traditional notions may still be valid. From the monkey's viewpoint, the discriminative trial starts with an observing response. Now, if the information is static during the prechoice response period all is well and the monkey need only use this information to respond appropriately. However, when the information is dynamic and may change prior to a response, which information controls behavior? If the monkey is a sequential operator, he initially may base his discriminatory response on the information first presented in order to determine its relevancy. For this information to be adequately evaluated, it may be necessary to ignore, disregard, or perhaps actively inhibit later information. If this initial information proves not to lead to consistent reinforcement, then the primate eventually must reverse the sequence, ignore the first information presented, and operate on the information available later in the prechoice response period. Sampling of information then becomes governed by the relevancy of the initially presented information. This sort of operational sequence has certain striking similarities to biological feedback schemes, such as those described by the TOTE mechanism suggested by Pribram (1971), and it is a generalized and highly efficient mechanism for evaluating dynamic informational input. If relevant information is presented during the early portion of the discrimination trial with later information actively inhibited, then as far as the primate is concerned, the only information preceding a choice response is the early information. By exclusion, this then necessarily makes this early information "functionally" closest to the choice response. Following this line of reasoning, the present data are not necessarily inconsistent with classical association theory, except that the monkey cannot easily be treated as a passive, inflexible system. Instead, a more realistic approach may be to recognize him as an active information gatherer, predisposed to one strategy but flexible enough to modify this strategy when it fails. The important consideration of this research therefore lies not so much in questioning the general position of association theory, but in refining its description of the manner in which association theory may operate when the monkey must encode and use dynamic information to guide its behavior.

POSTCHOICE RESPONSE PROCESSES

One thing appears clear from currently available data—reinforcement is not simply some physical event that signals the end of the discrimination process. Instead, the discrimination process appears to continue for some time following a choice response, for monkeys (Bartus, 1970; Bartus & Johnson, 1975; LeVere, 1967; LeVere & Bartus, 1973; Sheridan, Horel, & Meyer, 1962), humans (Dickerson & Ellis, 1965; Treichler, Hann, & Way, 1967; Treichler & Way, 1968), and rats (Enninger, 1953) all seem sensitive to postresponse events other than the occurrence or nonoccurrence of reinforcement. We now consider the question of how these events influence the acquisition of discriminatory behavior.

Taking a lead from certain physiological investigations, which suggest that postresponse trauma may disrupt normal memory functioning (see McGaugh &

Herz, 1972; Deustch, 1973), we propose that memory, or rather the appropriate functioning of memory, may depend on the character and nature of the informational events that follow a choice response. Our goal is not simply to reaffirm this position using nontraumatic stimulus events but to start from it and, by describing certain manipulations of postresponse stimulus information, provide some indication of the operation of memory in two-choice discrimination learning. In particular, our methodology has involved a procedure effectively used with monkeys by Sheridan *et al.* (1962), descriptively labeled "response-induced stimulus change" (RISC). In this procedure the discriminative cues are changed at the instant a choice response is completed. By noting the effects of particular changes on acquisition rate, we hoped to learn how the monkey used postresponse information and eventually to gain better insight concerning the mode of memory operation.

A study by LeVere (1967) helped demonstrate that the deficits resulting from postresponse stimulus information were most likely caused by some influence on memory processes. Most of the earlier postresponse results could be interpreted equally well using two opposing hypotheses. One of these assumed that secondary reinforcement was responsible for the effects of postresponse stimulus information on acquisition rate. It was contended that because of their simultaneous occurrence, an association occurred between the postresponse cue and primary reinforcement. Postresponse cue reversal, or in some cases persistence of the negative cue, would therefore retard acquisition rate caused by an inappropriate stimulus–reinforcement association. Persistence of the positive cue might be expected to increase acquisition rate. The other hypothesis accepted the traditional assumption that memory processes were most susceptible to external influence immediately following a response, and that postresponse information actually affected acquisition rate by interacting with these ongoing memory processes.

LeVere's (1967) study helped determine which of these two hypotheses was the more accurate by demonstrating that the stimulus information following a choice response need not be the information the individual is discriminating during the preresponse period in order for it to interfere with acquisition. This was accomplished by using irrelevant information (information different from the relevant information and unrelated to the reinforcement contingencies) as the postresponse stimuli. This information was never presented with the relevant preresponse information, and neither of the two irrelevant cues were consistently correlated or presented with the reinforcement. Given this situation, it would be extremely unlikely for secondary reinforcement to determine the effects of postresponse information. Yet, irrelevant postresponse information significantly interfered with acquisition. For this reason it is plausible that the presence of this irrelevant information during the time memory processes are presumably active interfered with proper storage of relevant information into memory.

Assuming that postresponse events do interfere with memory processes, how they interfere may indicate something about how memory operates. Two initial questions are therefore of some interest: Why does the subject continue to interact with stimulus information after the response, and how does this interaction influence memory? A followup to LeVere's (1967) study provided some suggestions.

Bartus (1970) attempted to provide greater detail regarding the effect of various types of postresponse information and also assessed whether postresponse cues might interact with encoding and remembering preresponse information. In this study, two types of preresponse stimulus information were presented, one involving only relevant information and the other involving relevant information obscured by being embedded within irrelevant information. During the postresponse period four different types of information were presented: relevant, irrelevant, relevant embedded within irrelevant, or no information, giving a total of eight treatments. If the processing of postresponse information significantly influenced the manner in which the preresponse information was encoded, an interaction might occur between acquisitions with the two types of preresponse information and the type of information following a choice response. Failure to observe any such interaction would suggest that these postresponse stimulus events influenced the discrimination process independent of how the preresponse information was encoded.

In order to equate the monkey's experience with the RISC procedure, they were run in balanced groups to a mean group criterion. Each of the eight treatments were represented in each of eight different groups. Training for all monkeys on any treatment always was terminated when the respective group reached a mean performance level of 85% correct responses on a given day. The major results of this study are shown in Fig. 8.6. It illustrates the relative effect of each treatment by depicting the mean percent achieved on each treatment at the time the balanced groups reached the group criterion. Despite a significant decrease in acquisition rate with embedded preresponse information, as well as significant differential postresponse effects, a significant interaction between the two preresponse treatments and the four postresponse information conditions was not substantiated. The most salient effect in this study was a consistent deficit associated with embedded preresponse information and this occurred regardless of the type of information following the response; it was not substantially lessened by postresponse presentation of relevant information alone, nor was it substantially worsened by postresponse presentation of irrelevant or embedded information. These data therefore indicate some degree of independence between the stimulus-related events occurring before a response and the stimulus events occurring after a response.

Although no interaction occurred between the pre- and postresponse treatments, significant differential effects of postresponse information were found. Presentation of relevant postresponse information did not facilitate learning,

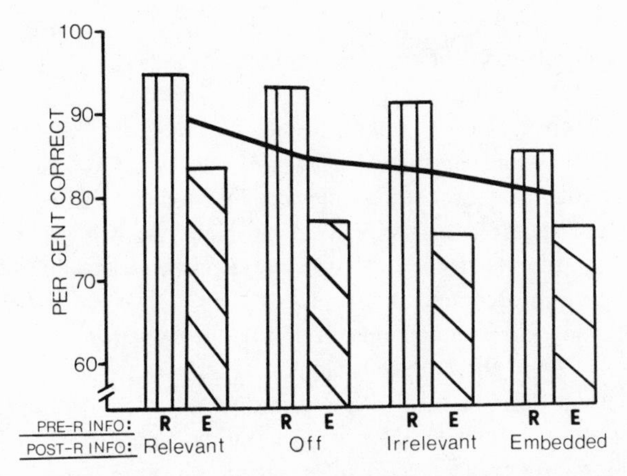

FIG. 8.6 Mean performance of all monkeys when each balanced group reached a mean criterion score of 85% correct responses within a single test session (see text for discussion of experimental design). The horizontal line running across the vertical bars illustrates the mean postresponse score, collapsed across the preresponse treatments. The deficit related to embedding the preresponse information was shown to be highly reliable (R versus E) when collapsed across the four postresponse conditions. Also, embedded postresponse information produced a reliable deficit, whereas postresponse irrelevant information approached significance. There was no deficit associated with the off postresponse condition, and the test for an interaction of pre- by postresponse information was not significant.

whereas presentation of irrelevant alone did not impair it (although a trend toward significance was observed). When relevant and irrelevant information were presented simultaneously (that is, embedded cue), however, a strong and reliable deficit did occur. These results replicated LeVere's (1967) finding that the most dramatic effect of postresponse stimulus information was interference with discrimination learning. A common finding therefore is that little or no facilitation typically is found with persistence of relevant information, while at the same time, presentation of irrelevant information substantially interferes with learning. Furthermore, the Bartus (1970) study demonstrated that this interference was greater and more reliable when the postresponse relevant information was embedded with irrelevant information than when irrelevant information was presented alone. This seemingly curious finding is discussed later.

If we are correct in assuming that postresponse stimuli are actively processed and that this processing can actually degrade the appropriate operation of memory, then it becomes of interest to determine whether or not the obtained effects depend on the occurrence of reinforcement. This information may shed light on the operation of memory as it relates to response outcomes and may provide the opportunity to examine directly whether secondary reinforcement somehow still can be operating to produce the effects of postresponse stimulus information. If it were demonstrated that the effects of postresponse stimulus information did not depend on the occurrence of reinforcement, it would be difficult to contend that secondary reinforcement was somehow responsible for approach tendencies to the incorrect discriminative cue. To evaluate the influence of reinforcement versus nonreinforcement in the effect of postresponse information, LeVere and Bartus (1973) trained Rhesus monkeys on discrimination problems that presented only relevant information during the trial. After each monkey mastered a discrimination problem, he was given one of three postcriterion treatments using the same preresponse discriminative stimuli. In one treatment, reinforcement continued to be delivered after each correct response, but now the preresponse relevant information was embedded in postresponse irrelevant information. On the other two treatments, the monkeys were shifted to a variable-ratio (VR) schedule of reinforcement, so that the majority of their correct responses would not be reinforced. In one of these latter two treatments only relevant information continued to be presented after the response, whereas in the other, the relevant information was embedded in irrelevant information.

Figure 8.7 shows that embedding the relevant information within irrelevant information has had absolutely no effect on the monkeys' performance, providing primary reinforcement has followed each correct response. However, when his correct choice responses were not consistently reinforced, embedding the relevant preresponse information did have a significant and detrimental effect on performance. This decrement was significant when compared to the control condition, where the monkeys were also shifted to a VR schedule but were presented with relevant stimulus information following choice responses. Although some of the deficit on the VR–embedded treatment can be simply attributed to the inconsistent delivery of reinforcement, a major portion of it depends on the fact that embedded stimulus information has followed these nonreinforced responses. When the above data are considered, a clearer picture of the influence of postresponse stimulus information emerges. Because the greatest effects have been obtained on nonreinforced trials, we can reaffirm that secondary reinforcement has not been responsible for these effects. However, if events important to memory are most active immediately following a response, the deficit resulting from postresponse irrelevant information may be related to its interference with memory processes associated with that trial's past events. Moreover, this interference does not occur equally well on all trials but is considerably greater on nonreinforced trials.

We therefore argued that in addition to whatever else it might do, reinforcement signaled the adequacy of the memory processes responsible for a choice response (LeVere & Bartus, 1973). In particular, the failure of reinforcement signals the inadequacy of the memory processes which precipitated the choice response and that the memory processes giving rise to the choice response must be modified or revised in order for improvement on the discrimination task to occur. If memory is revised following nonreinforced responses, this revision may

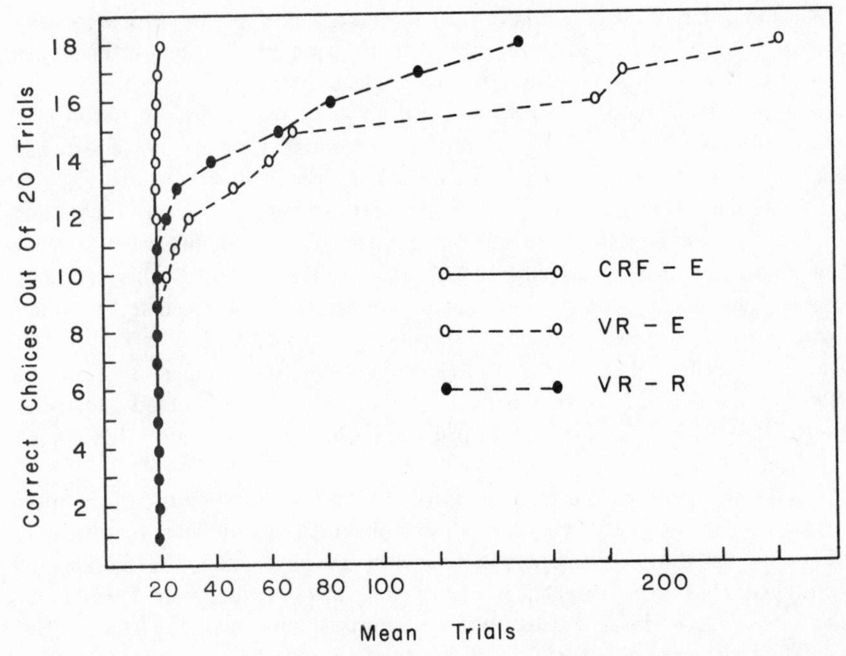

FIG. 8.7 Mean trials to re-attain criterion on a problem that was mastered with only relevant information presented throughout the trial. The data demonstrate the effect of presenting embedded postresponse information while the monkey is consistently reinforced for correct choice response (CRF–E), as opposed to inconsistently reinforced via a variable ratio schedule of reinforcement (VR–E). The simple effect of inconsistent reinforcement for correct responses is shown by continuing to present only relevant information throughout the trial after the VR schedule was introduced (VR–R).

make memory liable to (or perhaps the revision process may actively involve) the assimilation and utilization of available stimulus information. Obviously, if the stimuli present at this time actually represent misinformation, or information irrelevant to the task, learning should certainly suffer.

These results share striking similarities with data collected in a classical conditioning procedure by Wagner, Rudy, and Whitlow (1973). They found evidence that stimulus combinations that were unexpected and irrelevant to learning of the new conditioning problem impaired classical conditioning when presented after each trial. The authors convincingly argued that this impairment was caused by interference with a limited-capacity, time-dependent memory process, which they labeled "rehearsal." (For a more detailed discussion of these results, see chapter 9 in this volume.) Assuming that the nonoccurrence of reinforcement is somewhat unexpected to the monkey, our present results are directly analogous to those of Wagner *et al.* (1973). Because the two learning situations have many procedural differences, and yet similar interference occurs under somewhat similar circumstances, these data may point out a very basic process involved with the storage and modification of information in animal memory. More work in this area obviously needs to be done.

An unanswered question is why acquisition in the visual discrimination task suffers more when embedded information is presented after the choice response than when irrelevant information is presented alone. Intuitively, one expects greater deficits to occur with an irrelevant postresponse cue because it contains even less information than can be appropriately used to modify memory than does the embedded cue. If one could uncover the source of this apparent paradix, one might be able to define more precisely the manner by which postresponse information influenced revision of memory during learning. In an attempt to isolate the feature(s) most responsible for the greater disruptive effects of the embedded cue, Bartus and Johnson (1975) conducted a series of studies that varied the type of information which occurred after the choice response.

Because the embedded cue consists of relevant and irrelevant information presented simultaneously, one of the most obvious characteristics that make it different from the irrelevant cue is its greater size and complexity. Because of their greater size and complexity these cues may also be more salient, and saliency may contribute to disruption of postresponse memory. That is, the embedded cue may elicit greater visual sampling and therefore be more interfering with ongoing memory processes.

As an initial test to determine which critical features made the embedded cue disruptive, Bartus and Johnson trained two groups of monkeys with irrelevant postresponse information, differing only in terms of the size and complexity of the stimuli. The irrelevant stimulus for one treatment consisted of a single pair of irrelevant patterns. The other treatment used two separate pairs of irrelevant patterns that were superimposed. This latter, compound, irrelevant cue contained the same size and complexity characteristics as the embedded cue used in the previously described experiments. If sheer size or complexity of the embedded cue is primarily responsible for the greater effect on learning (or memory), greater impairment by the treatment with two superimposed irrelevant postresponse cues would be expected. The results of this study showed that no significant differences were found between learning with the single or compound irrelevant postresponse stimuli. Whatever difference did exist was actually opposite to the predicted direction.

Because quantity of stimulus information apparently was not the effective variable, the next possibility investigated was stimulus quality. One of the most apparent qualitative differences between embedded and irrelevant information is the presence of the relevant information within the embedded cue. Because the presence of the relevant information may make the embedded cue appear more similar to the relevant stimuli, the interfering effects of postresponse stimulus information may function in a manner similar to that reported for interfering stimuli presented during the retention interval in monkey and human short-term memory paradigms (D'Amato, 1973; Jarvik, Goldfarb, & Carley, 1969; Paivio & Bleasdale, 1974). In these situations, the more similar the interfering stimulus is to the relevant stimulus stored in memory, the greater the interference with retention (unless, of course, they are identical). The same similarity function may hold for irrelevant postresponse stimuli in the discrimination learning paradigm.

Two more sets of irrelevant stimuli, differing in their degree of similarity to the relevant stimuli, were designed. The relevant stimuli in this study consisted of pairs of parallel lines differing in orientation. The similar irrelevant stimuli also consisted of parallel lines, only differing from the two relevant stimuli in orientation. The dissimilar irrelevant stimuli consisted of a pair of geometric forms, bearing little or no resemblance to the parallel lines. The rationale for this test was that if cue similarity was responsible for the embedded cue's effectiveness, a reliable difference in rate of acquisition between the two irrelevant postresponse treatments would be obtained.

The results indicated that irrelevant information that differed only in orientation produced a significant impairment compared to a postresponse irrelevant cue that was markedly different from the relevant information. These data suggest that one reason the embedded cue effectively disrupts storage of postresponse memory is because it appears similar to but at the same time is a distorted representation of the relevant information on which the choice response has been based.

Bartus and Johnson proposed that the postresponse irrelevant information was most effective when it was confused with the encoded trace of the preresponse relevant stimuli. However, this notion requires the assumption that the monkey plays a more active role in processing and handling postresponse information than is usually assumed. For this reason a final test was designed to evaluate whether (a) the postresponse impairment was strictly the result of the embedded information passively interfering with the stored relevant information or (b) the monkey played a more active role in processing the information and therefore the eventual interference that occurred. Bartus and Johnson argued that if the monkey first learned which characteristics of the embedded cue were relevant and which characteristics were irrelevant, the disruptive effects of the embedded cue would be significantly reduced and this might negate the effects of the misinformation. That is, once the monkey learned to discriminate the relevant and irrelevant features of the embedded cue he would no longer confuse the postresponse embedded cue with the preresponse relevant information, and acquisition would not be impaired. If the monkey's role is more passive and the physical pre-requisites of the embedded stimuli remain, however, then interference should occur regardless of the monkey's prior experience.

To test these possibilities, Bartus and Johnson first trained monkeys on discrimination problems with embedded information throughout the trial. A prerequisite to learning such a problem would be to determine which features of the complex cue were relevant to solving the problem and which were irrelevant.

After each monkey mastered the discrimination problem with embedded pre- and postresponse information, it was trained on the reversal of that problem, with only relevant information presented during the preresponse period. For one of the two groups a new pair of irrelevant patterns was presented with the reversed cues, creating a "new" embedded postresponse cue; for another group the same "old" embedded postresponse cue used during initial learning was presented.

Learning of the reversal problem was substantially faster when the old embedded cue followed the response, than when new irrelevant information was used to make up the postresponse embedded cue. Once irrelevant features were differentiated from relevant features, the effectiveness of the embedded cue in disrupting memory storage was severely curtailed. However, when the embedded cue was comprised of new irrelevant information, the distinction between the relevant and irrelevant features again became unclear and a deficit resulted. The particular physical characteristics of the postresponse stimuli are therefore important to the extent to which they cause the monkey to confuse the relevant and irrelevant features of the pre- and postresponse stimuli. Once the monkey has learned to distinguish the particular relevant and irrelevant components, no confusion should occur and in fact no deficit results. Therefore, monkeys do seem to play an active role in determining whether irrelevant or inappropriate postresponse information interferes with acquisition.

Considered together, the emerging postresponse information data suggest that active and important processes continue to occur following a choice response. Apparently, in addition to whatever else it may do, reinforcement also serves as a signal that the information processed in that trial has been adequate and that no further stimulus processing is necessary. In contrast, if no reinforcement follows a choice response, the monkey becomes quite receptive to whatever postresponse information is available at that time.

Although there is yet relatively little evidence substantiating facilitation of acquisition caused by processing of postresponse stimulus information, there is no question that impairments in acquisition do result. Furthermore, these impairments are greatest when the irrelevant information is very similar to, but not identical to the relevant stimulus information. Little or no deficit results when the irrelevant is markedly different from the relevant information, or if the monkey first learns to discriminate the relevant from the irrelevant information. That is, these deficits depend on the degree to which the monkey perceives the irrelevant misinformation as being relevant to solving the discrimination task.

SOME CLOSURE AND SYNTHESIS

At a molar level, the research described may be related to discrimination learning by assuming that choice behavior is the overt expression of a correlation between the extrinsic stimulus information assimilated prior to a choice response and the intrinsic memories possessed by the individual. In the formal discrimination training situation, choice responses following this correlation can produce only one of two mutually exclusive outcomes–reinforcement or nonreinforcement. On the broadest level, one function of reinforcement is to indicate the appropriateness of this correlation, whereas nonreinforcement indicates that the correlation must somehow be modified if the discrimination problem is to be solved. To modify the correlation, the primate must operate conceptually on its interaction with the prechoice response extrinsic stimulus information or the intrinsic memory process associated with this information. The data summarized above suggest certain operational modes concerning how these two aspects of the correlation are modified with regard to the stimulus information that is available to the primate.

Consider first the question of how stimulus information presented before the response is processed and utilized. Our data suggest that at least with dynamic stimulus information, the subhuman primate approaches the problem as a sequential operator and bases his choice response on that information presented during the initial portion of the discrimination trial. Information available later in the discrimination trial is apparently ignored (perhaps actively inhibited) until the relevancy of the early information is established. It is important to note that this disregard for information later in the discrimination trial apparently transcends previous experience and reinforcement history. That is, information later in the discrimination trial is encoded only when it is apparent through the effects of "immediate" past nonreinforcement that the correlation between this information and memory is inappropriate. Because all monkeys eventually learn the discrimination problems with later occurring stimulus information, they must have abandoned (at least temporarily) their strategy of exclusively using the early occurring information. These data therefore support the idea that the monkey cannot be treated as a passive, inflexible system but is one who is able to modify predisposed strategies to meet the demands of the situation.

We turn now to the question of modifying an inappropriate correlation resulting from incorrect memory traces. Our data indicate that this involves certain interactions with available postresponse stimulus information. We suggest that reinforcement functions as a signal indicating the appropriateness of the correlation between memory and information encoding, and not simply as a physical drive reducer. Nonreinforcement, however, signals that the discrimination process must continue beyond the choice response so that what is stored in memory may be adjusted to meet the demands of the discrimination situation. It is with this interpretation, flavored strongly with Guthrie's (1940) classic notion of reinforcement that we feel the effects of irrelevant postresponse stimuli may be understood best. That is, when the response is reinforced the animal stops processing extrinsic information and therefore what is stored is immune from whatever irrelevant stimuli are present. However, when nonreinforcement follows the choice response the memory revision that occurs may in part be based upon the extrinsic stimuli available immediately following the incorrect choice.

If this information is irrelevant to the discrimination task, performance suffers. Appropriate memory traces cannot be generated if these memory traces are based on extrinsic stimulus information irrelevant to the discrimination task.

Here once again, however, the monkey demonstrates the active, dynamic nature of his interaction with the stimulus environment. If the irrelevant postresponse information is obviously different from the relevant information stored in memory before the response, then apparently no attempt is made to use it, or at least no deficit occurs. Even if the irrelevant information has the physical characteristics that can normally lead it to interfere with memory, no deficit occurs if the monkey is first trained to discriminate it from the relevant information needed to solve the discrimination task. Apparently, the monkey either actively samples the postresponse information for use in memory modification or, alternatively, actively inhibits its input when it is clear that information is irrelevant to solving the problem.

Why relevant postresponse stimulus information does not facilitate acquisition remains a question. It may be that this information adds little to what the primate covertly carries forward from the preresponse period. However, whether relevant postresponse information may be facilitative if the primate is discriminating compound or dynamic preresponse stimulus cues remains an important question yet to be answered. At present, there are only pilot suggestions and, as usual, further research is needed. However, if memory revision occurs, and if it can be interfered with by irrelevant information, then it is reasonable that in some situations it should also be facilitated by relevant postresponse information.

In summary, our experiments lead to the following conceptualization of information processing during discrimination learning; First, the primate selects particular aspects of the visual environment on the basis of the physical topography of his choice response and certain temporal strategies involving informational encoding and processing. Second, he executes a choice response based on the stimulus information and its correlation with mnemonic processes. If the choice response is reinforced, the monkey terminates the discrimination process, at least as far as using extrinsic stimulus information is concerned. If the choice response is not reinforced, the subject must revise or alter the correlation between memory and the available visual information. This may occur either by modification of the temporal encoding strategy or by the revision of specific memory storage, with the result that on subsequent trials a slightly modified correlation exists between the newly encoded information and the mnemonic trace. If this correlation leads to reinforcement, the monkey's interaction with extrinsic stimulus information stops and the correlation remains–if it does not lead to reinforcement, the monkey's interaction with available information continues and a new correlation is established.

One important question is whether or not the modification of the temporal aspects of preresponse information processing, on the one hand, and the revision of memory, on the other hand, may occur simultaneously, or are independent and sequential. If so, which precedes which also requires an answer. Moreover, whether or not the revision or modification of memory following nonreinforcement involves strengthening those aspects which provide information leading to a correct choice response or, as recently suggested by Estes (1973), the elimination of those aspects which provide misinformation is also a matter of considerable concern.

Despite the existence of these intriguing questions, a clearer picture of those processes responsible for discrimination learning nevertheless begins to emerge. This picture shows the primate bringing into the learning situation particular response strategies that bear strong influence on how the stimulus information is processed. More importantly, however, it also shows him actively interacting with this information throughout the trial, continuing to do so after the choice response, and being capable of modifying or discontinuing these strategies when they prove to be inappropriate. To be sure, this analysis of how the primate stores and utilizes stimulus information certainly is not the final word. Yet, if it does nothing more than demonstrate that discrimination learning involves something more than simply selecting the appropriate discriminative cue, then our purpose has been served—a purpose we hope to be useful.

9

The Dynamics of Episodic Processing in Pavlovian Conditioning

Jesse William Whitlow, Jr.

Rockefeller University

INTRODUCTION

The contributions to this volume attest to the fact that there has been a growing awareness over the past decade that animal learning involves something more than a collection of overt anticipatory and instrumental responses. Instead, a description of what is learned even in such traditional experimental paradigms as Pavlovian and Thorndikian conditioning seems to require a more abstract characterization than fits comfortably into the "response-oriented" approach of the past. At the moment, the most promising candidates for such an abstract characterization of what is learned are those that have adopted a "memory-oriented" approach (Estes, 1973), drawing on the well-articulated store of theoretical and experimental work in human memory.

Within this general orientation there are, of course, several research themes, and it may be helpful to indicate two areas of research currently pursued from a memory-oriented viewpoint that are related to the focus of the present chapter.

One such research theme is concerned with stimulus processing in nonarticulate organisms, particularly as this may be similar or dissimilar to human short-term memory. Chapters 4 by Roberts and Grant and 5 by Medin in this volume, for example, provide more detailed discussions of this approach.

Another research theme is concerned with providing a more adequate description of the associative product of learning episodes. Research on such topics as the transfer from Pavlovian conditioning settings to instrumental learning tasks (e.g., Overmier & Bull, 1970; Hearst & Peterson, 1973), the role of species-specific reactions in instrumental training (e.g., Bolles, 1972; Shettleworth,

1973), and the analysis of concurrently monitored multiple-response measures (e.g., Black & DeToledo, 1972; Schneiderman, 1972) encourage a recognition that response-oriented analyses of what is learned in Pavlovian or Thorndikian settings have appeared satisfactory only because traditional evaluations have been limited in scope, typically to a single response system. As investigators have expanded the range of test procedures, it has become less acceptable to view the associative product of a learning episode as simply a change in the strength of the tendency to engage in some observable response. Instead, an adequate theoretical analysis seems to require a construct, such as "expectancy," to represent what is learned.

The present chapter represents a third area of research, the theme of which may be characterized loosely as the attempt to determine relations between stimulus processing and expectancies. My concern with this problem has developed as a natural progression of research that I have undertaken in collaboration with A. R. Wagner and his colleagues and that has guided our selection of a memory-oriented approach to Pavlovian conditioning. Whereas a quantitative, process theory of Pavlovian conditioning as yet remains a long-range goal, I should like to describe here some of the phenomena that have encouraged us to adopt a particular memory-oriented theoretical framework. In so doing, I shall be able to illustrate the general point that attending to the dynamics of episodic processing promises to enhance the scope of our theories of Pavlovian conditioning. More specifically, the presentation serves to illustrate some of the integration of available data, insight into old problems, and new questions for study provided by a principle of episodic processing (e.g., Krane & Wagner, 1975; Wagner, 1976; Wagner & Terry, 1975): *The effectiveness of the CS to act as a retrieval cue for a US representation is increased to the extent that there is joint rehearsal of the CS and US representations and is attenuated to the degree that there is rehearsal of the CS representation alone.* As this principle involves such terms as "retrieval effectiveness" and "rehearsal," which are not part of the traditional vocabulary of animal conditioning, it may be helpful to begin with a brief review of the background of theory and data that have led to the proposition. The subsequent discussion of applications of the general principle in some recent research then appears, as is appropriate, part of a larger concern.

PROCESSING OF "EXPECTED" VERSUS "SURPRISING" UNCONDITIONED STIMULI

Reasons for introducing the concept of a rehearsal process as a critical mechanism in association formation in animal conditioning have been cogently set forth by Wagner (1976). There are, Wagner noted, several distinct classes of observations in Pavlovian conditioning showing differential processing of the US depending on whether it is well signaled by available cues, hence is "expected," or is poorly signaled by available cues, hence is "surprising." Wagner has shown the theoretical integration of these observations permitted by adopting a general theoretical framework from studies on human memory (e.g., Atkinson & Shiffrin, 1968) that emphasizes the role of a rehearsal process.

This framework distinguishes between a transient, limited-capacity short-term memory (STM) and a more permanent, relatively unlimited-capacity long-term memory (LTM). The condition necessary for relatively long-term retention is the transfer of information into LTM, and such transfer is assumed to occur only while the information is maintained, or rehearsed, in STM. To demonstrate long-term retention, it is necessary to retrieve information from LTM by presenting appropriate retrieval cues. Students of Pavlovian conditioning are usually interested in the retention of associative information, in particular with the degree to which the conditioned stimulus (CS) comes to act as a retrieval cue for a representation of the unconditioned stimulus (US). Therefore, our general concern can be expressed as that of determining how variation in the rehearsal potential of the CS and the US in Pavlovian conditioning episodes affects the capacity of the CS to serve as an adequate retrieval cue for a representation of the US.

To account for the observations on differential processing of "expected" versus "surprising" USs, the important additional assumption is that surprising USs are especially likely to engage the rehearsal process. Drawing on studies in rabbit eyelid conditioning, Wagner illustrated the way such an assumption within the general theoretical framework provides theoretical links among observations on blocking effects, retrograde interference effects, and the short-term availability of information about the US.

Blocking Effects

Granted the assumed relation between rehearsal and transfer of information into LTM, it is clear that if a surprising US is more likely to engage the rehearsal process, then a surprising US should more effectively promote conditioning. In support of this conclusion, Wagner pointed to data on "blocking" effects.

In the paradigmatic case, a CS, *X*, is reinforced or nonreinforced in compound with a second CS, *A*, that has been experimentally treated so as to yield different signal values for the *AX* compound. The principle observation from such an experiment is that the amount of conditioned responding to *X*, observed in testing subsequent to presentations of the *AX* compound, varies systematically with the treatment history of *A*. This pattern of data can be illustrated with the results of two studies described by Wagner (1976). In both studies subjects were conditioned to two cues, *A* and *B*, so as to train *A* highly and to train *B* weakly. The pretraining was followed by a phase in which the subjects in separate, matched groups received compound training with cue *A* (Group *AX*) or cue *B* (Group *BX*) in compound with a third cue, *X*. The change in the retrieval effectiveness of CS_X produced by compound training with either a strongly conditioned cue, *A,* or a weakly conditioned cue, *B,* was assessed in a final series of nonreinforced presentations of *X* alone.

In the first study *X* was introduced after pretraining with *A* and *B,* and subjects received reinforced presentations of either the *AX* or the *BX* compound. The important result was that there was less responding to *X* following *AX* reinforcements than following *BX* reinforcements. That is, the change in the retrieval effectiveness of cue *X* was less after *X* was paired with the US under conditions in which its occurrence was well signaled by other available cues (i.e., *A*) than after pairing *X* with the US under conditions in which the US was poorly signaled by other available cues (i.e., *B*). Comparing responding to *X* observed immediately before and after compound training showed a 5% increase in Group *AX* and a 35% increase in Group *BX.*

The complementary study involved pretraining *X* as well as *A* and *B* so as to strongly condition *X,* then administering nonreinforced presentations of *AX* or *BX.* In this case Group *AX* received *X* in compound with cues that strongly signaled an outcome (i.e., the US) discrepant from the occurrence of no-US, whereas Group *BX* received *X* in compound with cues that only weakly signaled an outcome discrepant from no-US. Under these conditions there was a considerably larger change in the retrieval effectiveness of *X* in Group *AX* than in Group *BX.* That is, there was more extinction of responding to *X* in the former than in the latter group.

Retrograde Interference Effects

Accepting standard ancillary assumptions about rehearsal, notably that it is a time-dependent, limited-capacity process, one should expect that if surprising USs are particularly likely to be rehearsed, then they should also serve as potent sources of interference for the processing of other, temporally proximate events. Substantial support for this argument is provided in a series of experiments reported by Wagner, Rudy, and Whitlow (1973).

Using a retroactive interference paradigm, in which conditioning episodes involving a previously untrained CS consistently paired with a US were followed closely in time by a posttrial episode (PTE), Wagner *et al.* (1973) predicted that when PTEs consisted of surprising CS–US episodes they would produce more retroactive interference than when expected CS–US episodes comprised the PTE. They further anticipated that the extent to which surprising PTEs resulted in interference effects would depend on the temporal interval between the target trials and the PTEs.

The basic procedure was to give subjects discrimination training between two cues, *A* and *B,* in which one cue, *A,* was consistently followed by reinforcement. whereas the other cue, *B,* was consistently followed by nonreinforcement. A high level of discriminative responding to *A* and *B* was established over a series of training sessions, so that at the end of this training it was reasonable to assume that the US was expected in the presence of *A* and the absence of the US was expected in the presence of *B*. Hence, it was also plausible to generate surprising episodes by presenting cue *A* not followed by presentation of the US, or, similarly, by presenting cue *B* followed by the US.

At this point, subjects received simple acquisition training to an untrained cue, *C*, chosen so as to be orthogonal to the dimension of difference between *A* and *B*. Each CS_C–US episode was followed by a PTE, which was, for some subjects, congruent with their prior discrimination training, that is, a CS_A–US or a CS_B–no-US episode. The PTE for other subjects was incongruent with their prior discrimination training, that is, a CS_A–no-US or a CS_B–US episode. These incongruent episodes were expected to be particularly successful in competing for the limited-capacity rehearsal process engaged by CS_C–US conditioning trials; subjects receiving incongruent PTEs were therefore expected to show a special impairment in the acquisition of responding to cue *C*.

This expectation was confirmed in several experiments. The acquisition of responding in groups that received incongruent PTEs was consistently at a lower level than that of groups that received congruent PTEs. Relatively specific evidence of the retroactive nature of the interference produced by incongruent PTEs was found in a study in which two new cues, *C* and *D*, were introduced in the second phase of training. The cue designated *C* was the target for interference, and each CS_C–US pairing was followed by a PTE. Trials involving the other cue, CS_D–US pairings, were never followed by a PTE. The two types of trials were intermixed in each training session in such fashion that a PTE was equally likely to precede trials with *C* and trials with *D*, thereby equating for any proactive influence of the PTEs.

The acquisition of responding to both CS_C and CS_D was monitored in two groups of subjects, one of which received congruent PTEs and the other of which received incongruent PTEs after each CS_C–US trial. The major results of this experiment were that only the incongruent PTEs served to interfere with the acquisition of responding and, more importantly, this interference was specific to the conditioning of cue *C*. Responding to *D* was equally strong in the two groups of subjects. If incongruent PTEs impaired conditioning via either a general disruption of performance or some proactive interference mechanism, one would expect to see depressed responding to *D* as well as to *C* in the group receiving incongruent PTEs. Apparently, however, the interference produced by incongruent PTEs was a selective, retroactive effect confined to the immediately prior conditioning episodes involving cue *C*.

It is also important to note that Wagner *et al.* (1973) found no apparent difference between the effect of incongruent PTEs involving a US presentation (CS_B–US) and that of incongruent PTEs involving no US presentation (CS_A–no-US). That is, the impaired conditioning produced with incongruent PTEs was not associated with either the presence or the absence of the US per se but was apparently caused by the surprisingness of the occurrence or nonoccurrence of the US. Therefore, there was no basis for attributing the detrimental effect of interpolated incongruent PTEs simply to transfer of conditioning effects from the PTE to CS_C (see, e.g., Ost, 1969).

An evaluation of the temporal range over which the retroactive effect was obtained concluded the set of studies. Because the rehearsal process is assumed to decay with time, the interference of an incongruent PTE with processing of CS_C–US episodes ought to diminish as the interval between target trial and PTE increases. In line with this prediction, Wagner *et al.* (1973) found that over a range of 3–300 sec there was clearly a temporal gradient of effectiveness, such that more interference was produced the closer in time incongruent PTEs followed target trials.

Assessment of STM

Finally, differential rehearsal of surprising versus expected USs should result in a greater short-term availability of information about surprising as compared to expected USs. Wagner pointed to a study by Terry and Wagner (1975) as providing a relatively direct assessment of the availability of US information in STM. Terry and Wagner used a procedure developed by Pavlov and refined by Konorski and Lawicka, termed a "preparatory-releaser" paradigm. The general strategy is to let one stimulus (the preparatory event) serve as a signal for the momentary conditions of reinforcement of a subsequent conditioning episode involving a second stimulus presentation (the releaser event).

The particular arrangement employed by Terry and Wagner was to correlate either the presence or the absence of a reinforcer in the conditioning episode with the prior presentation of an isolated US. Over the course of a training sequence, the CS in the conditioning episode (the releaser) was irregularly followed by the presence and absence of a reinforcer. Subjects could respond to the releaser in a manner appropriate to its momentary reinforcement condition only if they were able to discriminate in memory between a recent US presentation and the lack of such stimulation. By varying the interval between the preparatory stimulus and the releaser, Terry and Wagner were able to show that control of discriminative responding was consistent with the assumed decay of information about the preparatory stimulus from STM.

In order to compare the short-term availability of the US representation when the US was surprising as opposed to when it was expected, Terry and Wagner followed a strategy similar to that of Wagner *et al.* (1973) within the retroactive interference paradigm. That is, subjects were trained in separate sessions in a simple Pavlovian discrimination between two CSs, CS+ always terminating with the US, CS– never terminating with the US, so as to produce a high level of discrimination between the two cues.

Finally, the investigators arranged preparatory-releaser test sessions like the training sessions except that when the preparatory US occurred it was preceded by one of the independently trained CSs. Thus the preparatory stimulus could be a CS(+)US or a CS(−)US event. Terry and Wagner assumed that in the case of the CS(−)US event, the US would be relatively surprising. If a surprising US is more available in STM than is an expected US, this should be reflected in differences between the level of discriminative responding to the releasing stimulus after the two types of preparatory events. Using several types of test sessions, Terry and Wagner consistently found that discriminative performance following CS(−)US preparatory events was superior, albeit not dramatically so, to performance following CS(+)US preparatory events.

Figure 9.1 shows the results from two replications of test sessions in which CS(−)US, CS(+)US, CS− alone, and CS+ alone were evaluated as preparatory events. The performance measure is the magnitude of the deviation from the response level seen on trials in which no preparatory US had preceded the releasing stimulus ($\overline{\text{P}}$ episodes) observed following each of the preparatory events and signed + or − according to whether the deviation is or is not in the direction appropriate to the response level observed on trials in which the releaser has been preceded by a preparatory US.

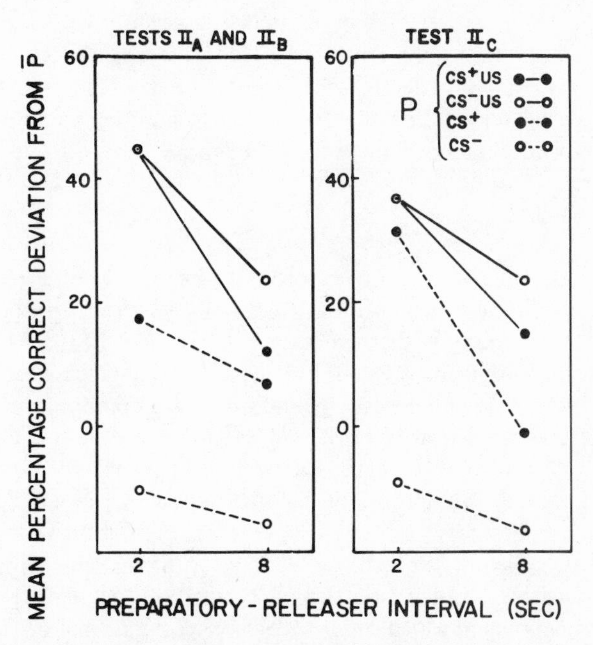

FIG. 9.1 Mean percentage conditioned responding to the releasing stimulus on P episodes during Test II, expressed as correct deviations from the percentage responding on $\overline{\text{P}}$ episodes. Plotted separately are the percentages from those P episodes involving only CS+ or CS– as preparatory stimuli, and those involving the compounds CS+–US and CS––US. The left panel presents the results from Test IIa and IIb in which the preparatory-releaser intervals were 2 and 8 sec, respectively. The right panel presents the results from Test IIc, containing both the 2- and 8-sec intervals. (From Terry & Wagner, 1975.)

As may be seen, performance following the surprising CS(−)US preparatory event was superior to that following the expected CS(+)US preparatory event, at least at the 8-sec interval (a ceiling effect prevented detecting differences at the 2-sec interval). This differential responding to the releasing stimulus after CS(+)US compared to CS(−)US was clearly not caused by preparatory effects of CS+ and CS−, as the individual CS presentations provided a marked contrast in their effect as preparatory events. When presented alone CS− acted to depress the distinctive mode of responding to the releasing stimulus below what was observed in those trials that occurred in the absence of a preparatory stimulus, whereas CS+ alone acted effectively as a substitute for the US.

Assuming that responding to the releasing stimulus was dependent on the immediate memory of the preparatory US, Terry and Wagner concluded that the short-term availability of the representation of a surprising US was greater than that of a representation of an expected US. In addition, Terry and Wagner noted that the results of tests involving CS+ and CS− alone were of considerable significance in themselves. The finding that CS+ acts as an effective substitute for the preparatory US is consistent with the theoretical description of the CS as a retrieval cue. That is, information about the US could appear to be in STM either as a result of a prior US presentation or as a result of presentation of a CS that retrieved a US representation from LTM. Moreover, they added, the fact that CS− was even less effective than the absence of a preparatory event suggested that "whatever CS− retrieved from LTM acted to diminish the representation of the UCS in immediate memory" (Terry & Wagner, 1975, p. 132).

Summary

These three classes of experiments, involving blocking effects, retrograde interference effects, and the short-term availability of US information, provide three independent demonstrations that processing of expected USs is different from the processing of surprising USs. Taken separately, of course, each type of observation is open to alternative theoretical interpretations other than the memory-oriented view described by Wagner (1976). What is especially attractive about Wagner's formulation is that it integrates all three classes of observations. That is, granted the assumptions of the theoretical framework described by Wagner, these studies may be viewed as converging lines of evidence for the proposition (e.g., Kamin, 1968; Wagner, 1971, 1976) that the US in Pavlovian conditioning initiates transient, poststimulation processing which is critical for association formation and which varies with the expectedness of the US.

The assumption that post-US processing is important for association formation represents an important departure from "response-oriented" approaches to theories of Pavlovian conditioning (e.g., Estes, 1955), which have ignored if not rejected outright such an assumption. I think this traditional stance concerning post-US processing has been governed largely by consideration of a single problem. It is therefore appropriate to turn to a reconsideration of that problem as the first example of the utility of the principle of episodic processing.

THE PROBLEM OF BACKWARD CONDITIONING

Traditionally, the major impetus for process theories of Pavlovian conditioning has been to account for the effects on conditioning of the interstimulus interval (ISI) between the CS and the US in a conditioning episode. The ISI function, as it is usually labeled, relating the effectiveness of conditioning episodes to the ISI between the CS and US, has been extensively investigated, and the general form of the function shows conditioning episodes to be less effective the longer the interval by which the CS antedates the US. There has admittedly been considerable debate about details of this general description, such as whether or not there is an optimal interval for conditioning, thereby making the ISI function nonmonotonic. Consideration of these details, however, has encountered a number of still unresolved interpretive difficulties and has not provided any significant distinctions among alternative theories.

What has been critical for theoretical accounts of the ISI function is the importance of the temporal order of the CS and US. Conditioning episodes in which the CS follows the US, or backward conditioning trials, have proved so ineffective in establishing an excitatory CS that the generally accepted view is that backward conditioning does not establish an association between the CS and the US (e.g., Anokhin, 1974; Kimble, 1961; Mackintosh, 1974). Indeed, several studies in which extended training was given with backward conditioning trials have produced evidence that asymptotically the CS in a backward procedure is associated with the absence of the US and has inhibitory response tendencies (e.g., Moscovitch & Lolordo, 1968; Siegel & Domjan, 1971).

If one accepts the conclusion that backward conditioning does not result in the formation of associations between the CS and the US, then a natural assumption to make about processing in Pavlovian conditioning is that such processing as is important for association formation is usually completed by the time of termination of the US. Within "response-oriented" theories, for example, it has frequently been assumed that the US serves only as a way of eliciting a response that is contiguous with the CS or the trace of the CS (e.g., Estes, 1955). The general acceptance of the assumption that associative processing is, if not instantaneous, at most a brief, discrete event attests to the parsimonious account of the ISI function such an assumption permits. It also attests to the lack of alternative data bases with which to evaluate a process theory.

If one assumes that post-US processing is not important for association formation, there is, of course, no reason to expect backward conditioning to generate an associative excitatory effect. The background research described above leads to the conclusion that associative processing is not a brief, discrete event, however. It is therefore natural to wonder why backward conditioning is apparently such an ineffective procedure for establishing excitatory response tendencies to the CS. Wagner and Terry (1975) have recently provided an interesting answer to this problem.

In reexamining the literature they noted that in fact there appeared to be a reproducible, excitatory effect of initial backward conditioning trials (Pavlov, 1928, p. 381; Razran, 1956; Switzer, 1930; Spooner & Kellogg, 1947), although after continued training such tendencies tended to disappear (Kamin, 1963; Smith, Coleman, & Gormezano, 1969). Other writers who have noted such transient excitatory effects have usually attributed them to nonassociative effects, such as "pseudo-conditioning" or "sensitization" (see, e.g., Mackintosh, 1974). Wagner and Terry, however, were attracted to an alternative interpretation of these excitatory effects as an associative product of the conditioning episodes.

An important characteristic of the backward conditioning procedure is that the US is not signaled by any explicit cue. This condition does not prevent the US from becoming expected, however. A substantial literature indicates that presentations of a US in the absence of an isolable cue results in the formation of an association between the contextual cues of the experimental situation and the US (e.g., Dweck & Wagner, 1970; Reberg, 1972; Rescorla & Wagner, 1972). Therefore, it is reasonable to assume that in the course of backward conditioning training the subject comes to expect the US to occur in the experimental context. According to the hypothesized relation between rehearsal potential and expectancy discussed above, the result of the development of these contextual associations should be attenuation of the rehearsal initiated by US presentations. Over the course of a training sequence of backward conditioning trials, it is reasonable to assume the US to be relatively surprising on the initial trials and thus to be quite effective in engaging the rehearsal process. Presentations of the CS placed shortly after the US would be likely to occur during the time of the processing necessary for association formation; hence the CS would pick up excitatory tendencies. With continued backward pairings, however, the subject would come to expect the US in the experimental context, so US rehearsal would be attenuated, and the CS would be more likely to occur in the absence of US rehearsal. Clearly, presentations of the CS in the absence of US rehearsal characterize extinction trials. Therefore, CS presentations occurring in the later phase of backward pairings might be expected to result in a loss of the initial excitatory tendencies.

This reasoning has led Wagner and Terry to the prediction that manipulations that insure the continued rehearsal of the US over a series of conditioning trials should result in more robust excitatory conditioning with backward pairings. A report by Heth and Rescorla (1973) of an experiment using the conditioned emotional response (CER) of rats to index association formation is in line with this prediction. Heth and Rescorla borrowed a design introduced by Mowrer and

Aiken (1954) involving an unusual set of conditioning parameters in that the duration of the US was greater than the duration of the CS. It was therefore possible to arrange a condition in which the CS occurred after the onset of the US but in which the US was maintained beyond the termination of the CS. The extended US presentation would be expected to maintain US processing during the time of occurrence of the CS and thus promote excitatory conditioning. In fact, Heth and Rescorla found in a subsequent test session that the CS in this condition did have reliable excitatory response tendencies.

Wagner and Terry (1975) obtained more direct support for the variable US-processing interpretation of backward conditioning. They evaluated this interpretation by comparing the effectiveness of two types of backward conditioning trials, each of which involved a distinctive CS. The backward pairings of the US with one cue, *C*, were always preceded by a cue that otherwise signaled the nonoccurrence of the US, thereby making the US surprising on backward conditioning trials. Trials in which the other cue, *D*, was presented in a backward pairing with the US were always preceded by a cue that otherwise signaled the occurrence of the US; hence the US on the backward conditioning trial was expected. Wagner and Terry anticipated that backward pairings would be more effective in the case of cue *C* than in the case of cue *D*.

Two groups of subjects received preliminary discrimination training with one cue, *A*, consistently followed by reinforcement, whereas the other cue, *B*, was consistently followed by nonreinforcement. The stimuli used as CSs, a 30 per sec train of clicks and a 3200-Hz pure tone, were 1100 msec in duration and, when reinforced, overlapped and terminated with a 100-msec, 4.5-mA shock applied to the infraorbital area of the eye. The designation of each stimulus as *A* or *B* was appropriately counterbalanced across different subgroups. As in the studies described earlier, this preliminary phase permitted the subsequent generation of CS–US combinations that could be considered either expected (i.e., CS_A–US) or surprising (i.e., CS_B–US).

Following this preliminary training, subjects in the backward conditioning group began to receive one backward conditioning trial each day embedded in the daily schedule of discrimination training trials. On some days, this backward conditioning trial involved pairing cue *C* with a surprising US. On these trials the stimulus sequence consisted of presenting cue *B*, which otherwise signaled nonreinforcement, so as to overlap and terminate with the US presentation, then 500 msec after US offset presenting cue *C* for 500 msec. On other days, the backward conditioning trial involved pairing cue *D* with an expected US. On these trials, the stimulus sequence consisted of cue *A*, which otherwise signaled reinforcement, followed by the US and presentation of cue *D* for 500 msec beginning 500 msec after US offset. At the end of each daily training session, subjects received a single, nonreinforced test trial to whichever cue, *C* or *D*, had been involved in the backward pairing in that session. A 20 per sec flashing light and a vibratory stimulus in contact with the subject's chest were employed as backward CSs, with the particular designation as *C* or *D* counterbalanced across different subjects.

The second group of subjects was intended to serve as a control for the possibility that surprising USs might produce more pseudo-conditioning or sensitization than expected USs. If there were such differential nonspecific effects of the two types of USs, it was conceivable that these effects might perseverate from the time of administration of the backward pairing to the time of the daily test, thereby producing differential amounts of test responding to cue *C* compared to cue *D*. In order to evaluate this possibility, subjects in the control group received a schedule similar to that of the backward conditioning group except that cue *C* and cue *D* were never presented in a backward pairing. That is, in place of the backward conditioning trial received by the backward conditioning subjects, control subjects received only a CS_B–US or a CS_A–US episode. In all other respects, such as the administration of a single test trial and the order of testing *C* and *D*, the treatment of the two groups was identical.

Figure 9.2 depicts for the two groups data from the test trials of this second phase of the experiment.[1] The data plotted in the figure represent the conditioned responding to cue *C* and cue *D* in the two groups of subjects, as well as the mean level of responding to the CS– in the discrimination trials that were maintained during this phase.

Two observations are of particular interest in Figure 9.2. First, it is clear that although the responding to cues *C* and *D* in the control group is appreciably greater than zero, there is no tendency for tests on those days in which a surprising CS–US episode has been embedded in the daily discrimination sequence to yield higher levels of responding than observed on tests on days in which an expected CS–US episode has been embedded in the normal sequence. Apparently, any nonspecific arousal properties of unexpected episodes did not perseverate until the time of the daily test. The responding seen to both cues presumably represented the combined effects of the net generalized excitation from the cues of discrimination training, *A* and *B*, and any residual nonassociative excitatory tendencies.

The observation of most interest, of course, is that in the backward conditioning group there is more responding to cue *C*, the CS that was placed subsequent to a surprising US, than there is to cue *D*, the CS placed subsequent to an expected US. In fact, the level of responding to cue *D* in the backward group is no different from that seen to either cue in the control group, suggesting that the backward pairing of cue *D* with an expected US did not produce any excitatory response tendencies. It may also be noted that responding to cue *C* in the backward conditioning group shows a steady increase across blocks of trials, suggesting that the backward pairings involving the surprising US have remained effective at least over the course of the training sequence.

[1] The author thanks A. R. Wagner and W. S. Terry for their generosity in making available to him the summary data used to draw Figure 9.2.

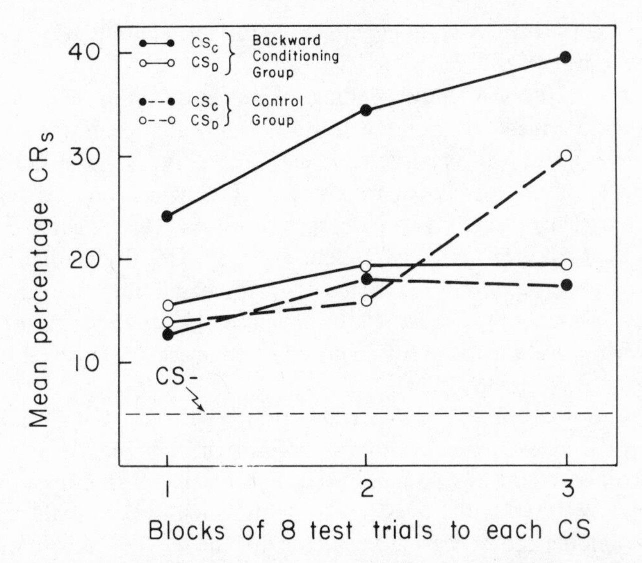

FIG. 9.2 Mean percentage eyelid responses over blocks of 16 test trials to each of CS_C and CS_D, plotted separately for the backward conditioning group and the control group. Test trials with CS_C occurred in sessions that involved presentation of a surprising US, those with CS_D in sessions that involved an expected US presentation. Also shown is the mean level of responding to the CS– used to generate a surprising US. (Based on data from Wagner & Terry, 1975.)

The results of this study clearly indicate that backward conditioning trials can be effective in establishing associations between the CS and the prior US. By showing that the effectiveness of such trials depends on the surprisingness of the US, the study provides further support for the proposal that variable rehearsal of the US is an important determinant of the development of those excitatory response tendencies in Pavlovian conditioning that reflect the effectiveness of the CS as a retrieval cue for US representations. One implication of the study is that the characteristics of backward conditioning are not a barrier to developing a process theory of Pavlovian conditioning that assumes variable poststimulation processing of the US. Of more general significance is the implication that complex associative effects, either incremental or decremental in nature, may result from the processing of post-US events.

"PREPAREDNESS" OR VARIABLE CONDITIONED STIMULUS PROCESSING

There are many ramifications of the proposal that post-US processing of events may either increment or decrement the retrieval effectiveness of those events. Here I should like to focus on one line of research that is a particularly intriguing example of the potentially complex effects of variable episodic processing on the retrieval effectiveness of CSs.

One of the most provocative findings documented in recent years is that gustatory CSs are made aversive much more readily than auditory or visual CSs when each is paired with toxicosis, whereas auditory or visual CSs are made aversive much more readily than gustatory CSs when each is paired with exteroceptive electric shock (e.g., Domjan & Wilson, 1972; Garcia & Koelling, 1966; Garcia, McGowan, Ervin, & Koelling, 1968). This finding has been taken to provide strong support for a concept of "preparedness" or "stimulus relevance" (e.g., Revusky & Garcia, 1970; Seligman, 1970) as an important determiner of the associability of events and as indicating the existence of a new form of learning (e.g., Seligman, 1970).

There is no question that the comparisons between taste aversions and auditory–visual aversions as a function of whether the US is a toxin or an electric shock have seriously undermined a traditional approach to measuring the salience of CSs. That is, the relative ease with which different stimuli come to control responding can be viewed as an index of the relative salience of the cues, such that more salient cues acquire conditioned response tendencies more easily than less salient cues. Because salience is assumed to be a property of the stimulus, moreover, the ordering of a set of stimuli observed when individual stimuli are paired with one US should not be changed when the stimuli are paired with another US. The failure to obtain such a trans-US ordering is a hallmark of the taste-aversion studies. Therefore, there is clearly a need to reconsider the assumption that ease of conditioning provides a reliable measure of salience. Yet such reconsideration does not mean that one must introduce a concept of "preparedness."

In the context of the principle of episodic processing, one alternative is to assume that the salience of a cue directly influences its processing, so that more salient stimuli are more likely to initiate rehearsal than less salient cues. The relation between salience and level of conditioning therefore depends on how variations in CS rehearsal affect the distributions of joint CS–US and CS alone rehearsal. For example, when US rehearsal is prolonged relative to CS rehearsal, then the increased processing afforded a more salient CS serves simply to increase the joint CS–US overlap. Hence increased CS salience should result in higher levels of conditioning.

Suppose, however, that US rehearsal is not prolonged relative to CS rehearsal, that is, CS rehearsal may extend beyond the termination of US rehearsal. Under this condition increasing CS salience increases the likelihood of CS-alone rehearsal. The net associative tendency, hence the level of conditioning, effected by a CS–US pairing depends on the combination of incremental effects from CS–US rehearsal and decremental effects from CS-alone rehearsal. Although predictions concerning the effects of salience on conditioning level in this case depend on specific assumptions concerning these incremental and decremental tendencies and their combination, it is clear that one possible prediction is that a less salient cue, i.e., one that initiates relatively weak rehearsal, is likely to yield a higher level of conditioning than a more salient cue that initiates more sustained rehearsal. Potentially this provides a boundary condition for the generalization that salience can be directly indexed by ease of conditioning. The generalization holds if US rehearsal is prolonged compared to CS rehearsal but may not be true if CS rehearsal extends beyond US rehearsal. Furthermore, this analysis potentially sets boundary conditions for comparing the ordering of the salience of CSs as indexed by conditioning with one US to the ordering as found through conditioning with another US. Such comparisons become meaningful only if the distributions of US rehearsal are similar in the two cases, so that, for example, rehearsal of either US is prolonged relative to CS rehearsal.

These speculations may seem a poor substitute for the notion of "preparedness." However, Krane and Wagner (1975) have reported a study that provides strong support for the general proposal that variations in CS processing may have a lot to do with the striking pattern of data found in the taste-aversion literature noted above.

As an alternative to "preparedness," Krane and Wagner (1975) have suggested that gustatory cues differ from auditory or visual CSs in terms of the duration of rehearsal initiated by the several cues. In particular, they have proposed that whereas the rehearsal duration of auditory or visual CSs is indeed usually short relative to the rehearsal initiated by an electric shock US, the rehearsal duration of gustatory CSs is prolonged relative to such a US. An implication of this assumption explored by Krane and Wagner (1975) is that the ISI function for gustatory cues may differ from that for auditory or visual cues.

That is, if CS rehearsal is brief relative to US rehearsal, then a maximal amount of joint CS–US rehearsal should arise if the CS and US occur close together in time. Increasing the delay between the two events reduces the availability of a CS representation, thereby reducing the potential for any joint CS–US rehearsal. There should be a simple decline in conditioning with increasing ISIs, therefore, in accord with the ISI function noted earlier, which is, I may add, based primarily on data involving auditory or visual CSs. However, what if CS rehearsal is prolonged relative to US rehearsal? The optimal interval is determined by what minimizes the detrimental effects of CS-alone rehearsal and maximizes the joint CS–US rehearsal, but the interval that satisfies these requirements is not necessarily one in which CS and US occur close together in time.

Indeed the results of the experiment reported by Krane and Wagner (1975) indicate that the optimal ISI for a gustatory cue paired with a brief US is considerably delayed. They employed electric shock as the US, which could be well controlled in its temporal relationship to a preceding gustatory stimulus or auditory–visual compound. Three groups of rats were allowed to consume water from a drinking tube for 20 sec, each lick producing either an auditory–visual signal in one case, or contact with saccharin flavor in the second case, or no explicit cue in the third case. At the end of the CS exposure period, the drinking tube was made inaccessible and a third of each group was presented with a brief, intense footshock, either 5 sec, 30 sec, or 210 sec following termination of the CS period. To evaluate the specific aversion acquired to the gustatory or auditory–visual stimuli following these conditions of pairing with shock, all subjects were eventually tested for their consumption of water when accompanied by the flavor or exteroceptive stimuli as compared to water alone.

Figure 9.3 summarizes the major results of the experiment. At each delay interval represented in the figure are depicted the suppression scores of subjects trained with a light–tone compound when tested with that compound and the scores of subjects trained with a saccharin flavor when tested with the flavor. In addition, the mean level of suppression to the light–tone compound and the saccharin flavor observed in testing subjects that have not received an explicit cue during conditioning are depicted above the point marked "PC" on the abscissa. These latter points provide a baseline with which to assess the amount of conditioned suppression to each CS.

As may be seen, there is a monotonic decline in conditioning to the light–tone compound as the ISI increases from 5 to 210 sec. In contrast, the level of conditioning apparent with the saccharin CS was greater following a delay of 30 or 210 sec than at the briefer delay of 5 sec. Furthermore, although the light–tone compound appeared to be the more strongly conditioned cue after training with a 5-sec ISI, the saccharin flavor appeared to be the more strongly conditioned cue after training with the longer ISIs.

Krane and Wagner noted one implication of their pattern of results that is methodological. In the preparedness literature, they argued, when electric shock has been used as US with either a gustatory or an exteroceptive CS, there has generally been a close temporal contiguity arranged between CS and US (e.g., Garcia *et al.,* 1968). When a toxin has been employed as US, again with either a gustatory or an exteroceptive CS, there has generally been an appreciable delay between exposure to the CS and application of the toxin necessitated by the imposition of an injective procedure or an appreciable delay between the subject's exposure to the CS and the onset of discomfort due to the time of development of the malaise (e.g., Garcia & Koelling, 1966). Such a confounding of CS–US delay with the nature of the US is obviously relevant given Krane's and Wagner's pattern of data.

The more fundamental point raised by Krane and Wagner is that an uncritical reception of the notion of "preparedness" tends to obscure the real significance of both their own data and those of others noted above on taste aversions–that "the critical question [raised by such data] is whether or not we can develop a *theory* that will allow us to deduce the observed laws [of conditioning], including those which involve interaction effects. [p. 888]."

Indeed, this research may be viewed as simply the most notable example of a larger class of observations that includes the interacting effects on conditioning of interstimulus interval and CS modality (Krane & Wagner, 1975), stimulus

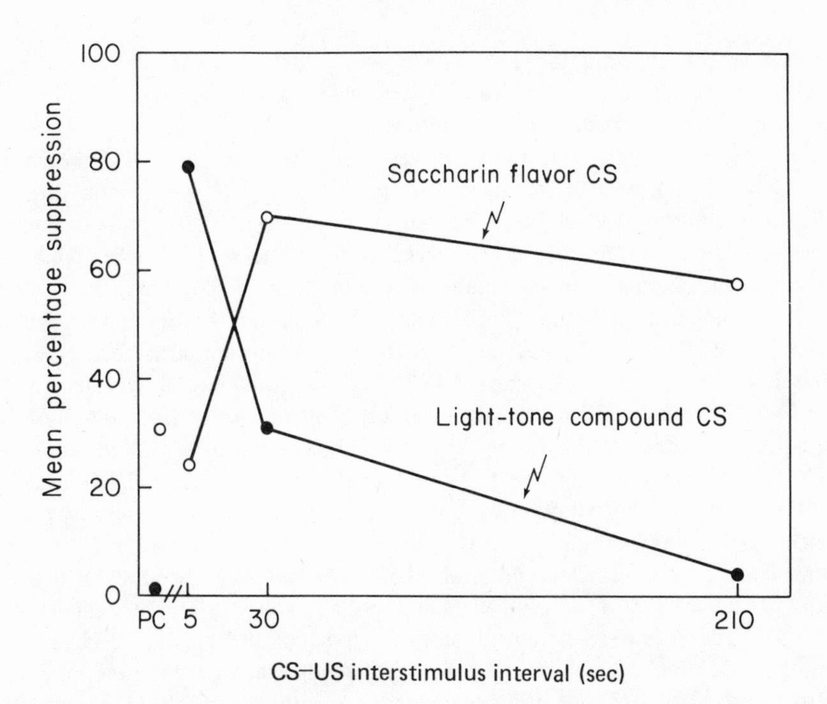

FIG. 9.3 Mean percentage suppression of licking in test sessions involving a cue previously paired with electric shock, in separate groups that had received either a light–tone compound or a saccharin flavor followed by shock after 5, 30, or 210 sec. The points above "PC" on the abscissa indicate the mean percentage suppression of licking in tests with the light–tone compound and the saccharin flavor for subjects that had received a shock in the absence of any explicit cue. (Based on data from Krane & Wagner, 1975.)

complexity (e.g., Mayer & Ross, 1969), and alpha response level (e.g., Martin & Levey, 1969). Whether or not the theoretical framework described in the present chapter can meet the challenge posed by such observations is a matter for future research. What I have tried to show is that a renewed interest in process theories of conditioning freed of some of the traditional limitations we have placed on our conceptions of episodic processing at least holds promise for enriching and enlarging our theoretical understanding of Pavlovian conditioning.

REINSTATEMENT AND VARIABILITY OF CONDITIONED STIMULUS PROCESSING

As a final example of the sort of complexities of conditioning that provoke theoretical consideration of the dynamics of episodic processing, I should like to share some observations from work still in progress. The empirical reproducibility of these data is well established, although their interpretation is not yet clear. However, there are some interesting similarities between this research and the studies both by Wagner and Terry (1975) and by Krane and Wagner (1975) that have been discussed. Pointing to such similarities therefore serves to illustrate possible extensions of those prior analyses.

The research to be considered (Donegan, Whitlow, & Wagner, 1975) began by asking whether the effectiveness of forward, CS–US pairings in simple acquisition of conditioning could be enhanced by a brief, posttrial re-presentation of the CS. What prompted this question was a report that such a brief re-presentation, or reinstatement, of a compound CS facilitated conditioning to an element of the compound. Gray and Appignanesi (1973) used the conditioned emotional response (CER) of rats to evaluate the effect on conditioning to *X* of reinstateing an *AX* compound after each *AX*–US pairing in a modified version of the "blocking effects" design described earlier. That is, cue *A* was pretrained to signal the occurrence of the US; hence the conditioning to cue *X* produced by *AX*–US pairings was expected to be relatively weak. This was indeed the case. The novel result of Gray and Appignanesi's procedure, however, was that reinstating the *AX* compound significantly enhanced conditioning to *X*. The immediate concern of Donegan *et al.* (1975) was with whether this facilitating effect of reinstatement was linked to the special conditions of a blocking study, as Gray and Appignanesi argued in their report of the phenomenon, or was another example of a more general phenomenon, that is, the conditions for efficacious backward conditioning, congruent with the principle of episodic processing. They therefore modified Gray and Appignanesi's procedure to examine reinstatement in simple CER acquisition training to a single isolable CS.

Rats were trained to bar press on a VI schedule. After VI training, each subject received at least two nonreinforced preexposures to a CS (either a tone or a diffuse increase in illumination) and then a series of pairings of the CS with a .5-sec, 1-mA foot shock, one trial per day. The experimental comparison involved two treatments, one a conventional conditioning sequence in which the CS was 3 min in duration and terminated with the shock US, the other a replica of the conventional sequence with the addition of a 500-msec reinstatement of the CS 10 sec after the US. The response measure employed was percentage suppression, that is, the percentage by which bar pressing was reduced in the presence of the 3-min CS as compared to the preceding 3-min period.

A general summary of results from this research is that under these conditions reinstatement has more complex effects than would be anticipated from Gray and Appignanesi's study. Donegan *et al.* (1975) did not find an overall effect of reinstatement. That is, the mean level of conditioning averaged across all subjects that received reinstatement was not different from that averaged across all subjects that did not receive reinstatement. Nonetheless, the reinstatement manipulation was far from being ineffective.

The phenomenon that is best substantiated in the research of Donegan and co-workers is that the effect of the reinstatement manipulation was either incremental or decremental, depending on subjects' initial responsiveness to the

CS. This pattern can be illustrated with data from a study involving 54 subjects, trained with a tone CS.

Figure 9.4 represents the frequency distribution from this study of the mean percentage suppression scores over two habituation trials and the CS presentation on the first conditioning trial, that is, over three presentations prior to the first shock. As may be seen, some subjects suppressed responding, others accelerated responding (as indexed by negative suppression scores), with the modal tendency being no effect.

To evaluate the relation between initial responsiveness and the effect of reinstatement, Donegan *et al.* (1975) made a simple distinction between low-responsive subjects and high-responsive subjects: Low-responsive subjects were that half of any treatment group which showed the closest to no effect of the CS on bar pressing prior to conditioning. High-responsive subjects were then the remaining half of the treatment group and showed more extreme changes in bar pressing when presented with the CS prior to conditioning. More exactly, as replications of their studies were run at different times and with some variation in the subjects, such as differences in age, the classification of high- versus low-responsive was made within each replication. The bars above the histogram in Fig. 9.4 show the actual ranges of suppression of subjects consequently included in the low-responsive category (stippled bar) and in the high-responsive category (solid bars) over all replications of this study.

For subjects that received CS–US pairings and no reinstatement, this responsiveness classification was a good predictor of the subsequent level of conditioned responding. For example, a regression analysis in which responsiveness was a categorical independent variable and the mean percentage suppression across Days 2 to 5 of conditioning was the dependent variable yielded an $r = .574$, $p < .005$. In other words, high-responsive subjects tended to show high levels of conditioned responding, whereas low-responsive subjects tended to show low levels of conditioned responding.

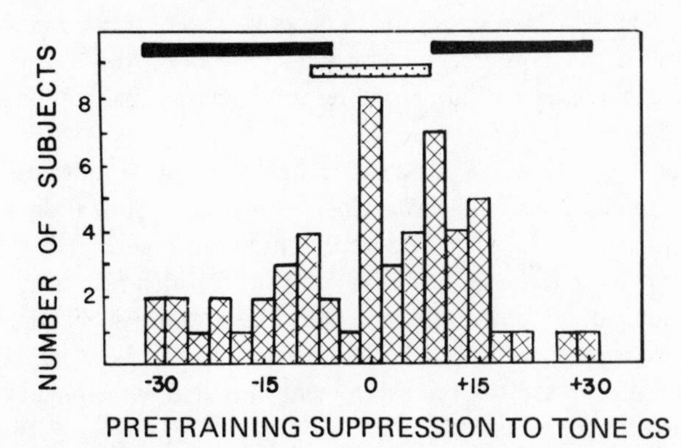

FIG. 9.4 Frequency histogram showing the distribution of pre-training suppression scores to a tonal CS. (From Donegan *et al.*, 1975.)

Reinstating the CS, however, acted to decouple initial responsiveness and conditioned responding. For subjects that received reinstatement, the correlation between responsiveness classification and percentage suppression across Days 2 to 5 of conditioning dropped to a nonsignificant level, $r = .153, p > .10$.

Figure 9.5a depicts the pattern of data reflected in these regression analyses. The figure indicates the course of acquisition of conditioned suppression for high- and low-responsive subjects in groups that have not received reinstatement and for the two types of subjects in groups that have received reinstatement. As may be seen, in the no-reinstatement conditions there is a substantial separation between the mean level of suppression seen in high- as compared to low-responsive subjects. With reinstatement, however, the mean levels of suppression in high- as compared to low-responsive subjects were very similar. More precisely, the effect of reinstatement was to increase the level of suppression shown by low-responsive subjects but to decrease slightly the level of suppression shown by high-responsive subjects.

Before I suggest possible interpretations of this phenomenon, I should describe some further observations made by Donegan *et al.* (1975) concerning the effect of reinstatement. There is little precedent in animal learning for attaching much theoretical significance to individual differences, and Donegan *et al.* (1975) have been appropriately concerned about possible adventitious features of the pattern depicted in Figure 9.5a. That the pattern is not peculiar to the tonal stimulus chosen as a CS is indicated by the results of a set of studies, involving a total of 62 subjects, in which a change in illumination served as the CS. The major difference between the visual CS and the auditory CS was found in the distribution of initial responsiveness. The visual CS tended overall to produce a slight suppression of responding, so that subjects classified as high-responsive in terms of the size of the deviation of their reaction to the CS from no effect tended to show an initial suppression of responding. Nevertheless, the high- vs low-responsiveness classification remained a good predictor of the subsequent level of suppression across Days 2 to 5 for subjects that were trained in the absence of reinstatement, as substantiated by an $r = .73, p < .005$. And reinstatement acted to remove the predictive utility of the responsiveness classification, indicated by an $r = .22, p > .10$.

Figure 9.5b depicts the course of acquisition of conditioned responding observed in these studies with visual CSs. Compared to similar data for tone CSs depicted in Figure 9.5a, the data represented in Figure 9.5b show a more pronounced decremental effect of reinstatement for high-responsive subjects and a less pronounced incremental effect for low-responsive subjects. The more important observation, however, is that the interaction between responsiveness classification and the effect of reinstatement obtained with tonal CSs has clearly been reproduced with visual CSs.

Of course, the observations depicted in Figure 9.5 represent the effects of reinstatement as observed over the course of several conditioning episodes. An

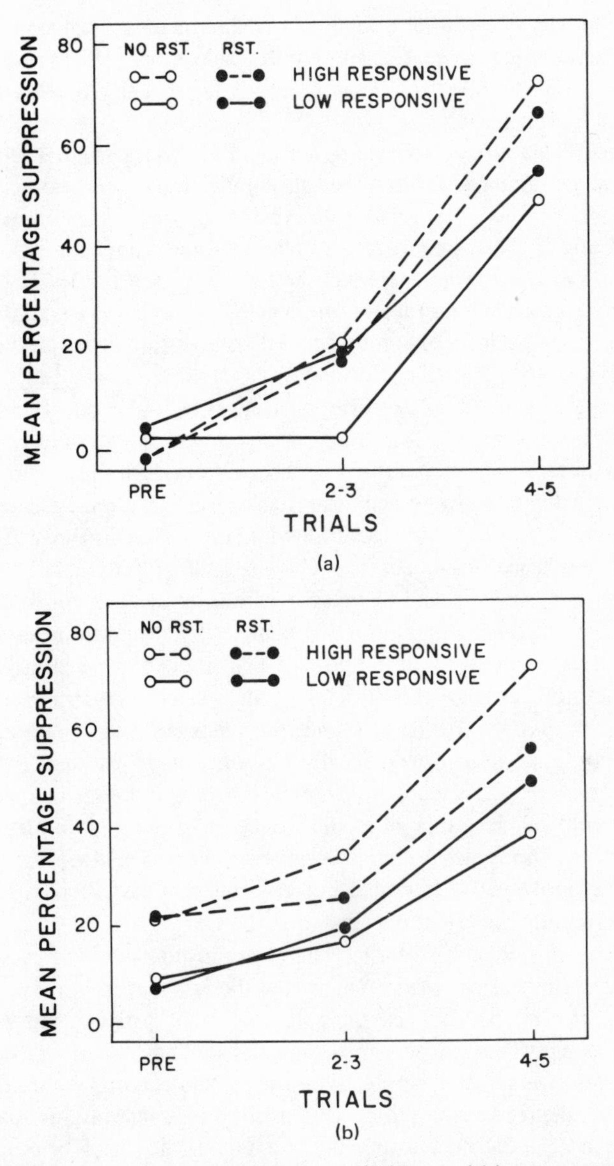

FIG. 9.5 (a) Mean percentage suppression scores during acquisition to a tone CS of groups receiving reinstatement or no-reinstatement, plotted separately for High-responsive and Low-responsive subjects. The leftmost points in each graph indicate the mean level of pretraining suppression to the CS in each of the four groups. (From Donegan *et al.,* 1975.) (b) Mean percentage suppression scores during acquisition training to a visual CS, depicted as in fig. 9.5a. (From Donegan *et al.,* 1975.)

indication that the differential effects of reinstatement are not simply multiple-trial phenomena is that in several studies Donegan *et al.* (1975) found differential effects of reinstatement on Trial 2, that is, after a single conditioning trial was followed by reinstating the CS.

The pattern of data depicted in Fig. 9.5a and 9.5b suggests that reinstatement acted merely to reduce the difference in conditioning level between high- and low-responsive subjects. This general description of the effects of reinstatement was also found to be inappropriate. Under some conditions Donegan and co-workers observed that reinstatement actually reversed the level of conditioning in high- versus low-responsive groups, such that low-responsive groups showed a higher mean level of conditioning than high-responsive groups.

A brief review of the pattern of data is this. Donegan *et al.* (1975) segregated subjects on the basis of an index of responsiveness to tonal and visual CSs. In the absence of reinstatement, this responsiveness classification was positively correlated with the level of conditioning observed during training with the CS on which the classification was based. Reinstating the CS eliminated this positive correlation, and it did so in the following interesting manner. Relative to conditioning levels seen in its absence, CS reinstatement facilitated conditioning in low-responsive subjects and depressed conditioning in high-responsive subjects. Furthermore, the manipulation was sufficiently strong to produce differential effects after a single conditioning trial and under some conditions even to reverse the relative levels of conditioning of high- versus low-responsive subjects.

The complex pattern of data found by Donegan and co-workers clearly indicates that Gray and Appignanesi's account of reinstatement is at best incomplete. However, it should be acknowledged that these data pose a challenge to theoretical accounts of conditioning in general, not just Gray and Appignanesi's. Certainly the principle of episodic processing does not lead one to anticipate that individual differences in responsiveness have such striking importance in determining the effect of reinstating the CS.

Nonetheless, an indication of the general applicability of the present framework is that it can suggest ways to approach the reinstatement effects found by Donegan *et al.* (1975). For example, a reasonable assumption is that the responsiveness classification described above is an index of the salience of the CS. High-responsive subjects would therefore be subjects for whom the CS was an especially salient event. Granted this assumption, one interpretation of the Donegan *et al.* (1975) data would be as a parallel to the Krane and Wagner (1975) study, in which reinstatement, like a gustatory CS, would be viewed as generating a relatively prolonged amount of total CS processing. The extent to which reinstatement prolonged CS rehearsal would, of course, be greater for high-responsive than for low-responsive subjects. Of the sets of possible rehearsal distributions in high- and low-responsive subjects, the joint realization of two is clearly relevant to the present discussion. These distributions are: (1) CS rehearsal in the former subjects is sufficiently prolonged relative to US rehearsal that the detrimental effects of CS-alone rehearsal outweigh any benefits of further CS–US rehearsal, thereby producing impaired conditioning in high-responsive subjects; and (2) CS rehearsal in low-responsive subjects is, by comparison, so attenuated that the benefit of additional CS–US rehearsal permitted by reinstatement offsets any decremental effects of CS-alone rehearsal, thereby producing facilitation in these subjects. To speak loosely in terms of processing effects in this analysis, reinstatement may be said to increase the salience of the CS, but to such an extent for high-responsive subjects that the relation between salience and level of conditioning has broken down.

An alternative interpretation of the Donegan *et al.* (1975) data that is also compatible with present framework is as a parallel to the Wagner and Terry (1975) study on backward conditioning. The sequence of stimulation on reinstatement trials, for example, is similar to that used by Wagner and Terry on backward conditioning trials, that is, a CS followed by a US which in turn is followed by a CS. One may still assume that the responsiveness classification indexes CS salience. In this interpretation, however, the important effect of salience is in terms of the rate at which the CS acquires signal value. Moreover, noting the positive correlation between responsiveness classification and level of conditioning when reinstatement is omitted, one may infer that a more salient CS comes to signal the US more quickly than a less salient CS. This permits a further analogy between the studies by Donegan and co-workers and by Wagner and Terry—that reinstatement is a backward, US–CS pairing which occurs after an expected US for high-responsive subjects but after a relatively surprising US for low-responsive subjects. Of course, Wagner and Terry did not find a reduction in the strength of a CS following an expected US, as would be indicated by the impairment Donegan *et al.* (1975) observed in high-responsive subjects. However, Wagner and Terry assumed that such a reduction did normally occur in more traditional backward conditioning studies; in their case, the CS presented after an expected US had no detectable strength against which any loss in strength could be assessed. It may be noted that this interpretation is not incompatible with the previous one, and that either or both can be applied to the Donegan *et al.* (1975) data.

These interpretations are admittedly speculative. The point I have tried to illustrate is that a flexible conception of the dynamics of episodic processing holds promise for the analysis and understanding of the sort of complexities Donegan and co-workers have found in their research on reinstatement. At the same time, I think this discussion shows how the principle of episodic processing presented earlier points to researchable questions about the relation between characteristics of episodic processing, even at the level of individual subjects, and the net associative effect of a conditioning episode.

CONCLUSION

The response-oriented tradition of psychology has often viewed Pavlovian conditioning as a singular paradigm in which the fundamental laws of learning can be disentangled with relative ease. In keeping with this view, the range of empirical phenomena obtained in Pavlovian conditioning situations seemed to consist of a small set of functional relationships that could be encompassed by corollaries to a principle of association formation stressing the temporal contiguity of stimulus and response.

The memory-oriented viewpoint of this chapter contrasts with this tradition in two ways. The first difference is that the present viewpoint considers Pavlovian conditioning only one of the laboratory preparations that may comment on the process of associative learning instead of necessarily being one of the basic "types of learning." The advantages of Pavlovian conditioning as a preparation are in its apparent skeletal simplicity and the relatively precise controls it allows over the subject's experiences, in the wealth of parametric data already available for particular situations, and of course in its suitability for investigation with nonarticulate organisms. However, one may expect there to be mutual benefits from theoretical exchanges between psychologists concerned with other preparations also presumed to comment on the process of associative learning. The present chapter is witness to the benefit to theories of Pavlovian conditioning of drawing on theoretical developments in the study of human memory. An indication that the exchange need not be asymmetrical is the analysis of paired associate learning by Rudy (1974).

The second difference is the stress on the temporal contiguity not of CS and US presentations but of the processing initiated by such presentations. What makes this change of emphasis theoretically significant is that it provides a natural framework in which to consider variable processing in Pavlovian conditioning, and the notion of variable processing appears to represent a significant theoretical advance. By combining the assumption that expected USs are less likely than surprising USs to engage a rehearsal process with a standard set of assumptions concerning STM and LTM, Wagner (1976) has been able to integrate observations on (a) blocking effects in compound conditioning with expected USs, (b) the greater retrograde interference effect of surprising vs expected USs, and (c) the reduced short-term availability of information about expected vs surprising US preparatory stimuli. To this list, the present chapter appends the acquisition of responding with backward pairings of a CS and a surprising US.

This chapter has in addition considered variable processing of the Pavlovian CS. The novel assumption has been that a more salient CS has greater rehearsal potential than a less salient one. The chapter has described an extension of Wagner's (1976) general framework to provide a theoretical account of the effect of differences in CS salience, whether arising from variation in individuals' responsiveness or from different CS modalities and to suggest that such an account may not only unravel some of the complexities of reinstatement and taste-aversion effects but also to integrate these observations with a larger body of data. Such a theoretical integration is one measure of the utility of a memory-oriented approach to conditioning.

ACKNOWLEDGMENTS

The research described in this paper was conducted while the author was a graduate fellow at Yale University and was supported in part by National Science Foundation Grants GB-30299X and BMS74-20521 to Allan R. Wagner. Preparation of the manuscript was supported in part by National Institute of Mental Health Grant MH16100.

References

Amsel, A. Frustrative nonreward in partial reinforcement and discrimination learning: Some recent history and a theoretical extension. *Psychological Review,* 1962, **69**, 306–308.

Anderson, J. R., & Bower, G. H. *Human associative memory.* Washington, D.C.: Winston, 1973.

Anokhin, P. K. *Biology and neurophysiology of the conditioned reflex and its role in adaptive behavior.* New York: Pergamon Press, 1974.

Atkinson, R. C., & Shiffrin, R. M. Human memory: A proposed system and its control processes. In K. W. Spence & J. T. Spence (Eds.), *The psychology of learning and motivation.* Vol. 2. New York: Academic Press, 1968.

Attneave, F. Symmetry, information and memory for patterns *American Journal of Psychology,* 1955, **68**, 209–222.

Bailey, C. J. The effectiveness of drives as cues. *Journal of Comparative & Physiological Psychology,* 1955, **48**, 183–187.

Barbizet, J. *Human memory and its pathology.* San Francisco: Freeman, 1970.

Barnett, S. A. Experiments in "neophobia" in wild and laboratory rats. *British Journal of Psychology,* 1958, **49**, 195–201.

Bartus, R. T. The influence of post-choice response stimulus information during two-choice discrimination learning. Unpublished master's thesis, North Carolina State University, Raleigh, 1970.

Bartus, R. T., & Johnson, H. R. Unpublished manuscript, 1975.

Behar, I. Analysis of object-alternation learning in rhesus monkeys. *Journal of Comparative & Physiological Psychology,* 1961, **54**, 539–542.

Bessemer, D. W., & Stollnitz, F. Retention of discriminations and an analysis of learning set. In A. M. Schrier & F. Stollnitz (Eds.), *Behavior of nonhuman primates.* Vol. 4. New York: Academic Press, 1971.

Bjork, R. A. Repetition and rehearsal mechanisms in models of short-term memory. In D. A. Norman (Ed.), *Models of human memory.* New York: Academic Press, 1970.

Bjork, R. A., & Abramowitz, R. L. The optimality and commutativity of successive interpresentation intervals in short-term memory. Paper presented at the meeting of the Midwest Psychological Association, Chicago, May 1968.

Black, A. H., & deToledo, L. The relationship among classically conditioned responses: Heart rate and skeletal behavior. In A. H. Black & W. F. Prokasy (Eds.), *Classical conditioning.* Vol. 2. New York: Appleton-Century-Crofts, 1972.

Bliss, D. K. Disassociated Learning and state-dependent retention induced by pentobarbitol in Rhesus monkeys. *Journal of Comparative & Physiological Psychology,* 1973, **84,** 149–161.

Blum, G. S., Graef, J. R., Hauenstein, L. S., & Passini, F. T. Distinctive mental contexts in long-term memory. *International Journal of Clinical & Experimental Hypnosis,* 1971, **19,** 117–133.

Bolles, R. C. Reinforcement, expectancy, and learning. *Psychological Review,* 1972, **79,** 394–409.

Bolles, R. C. Learning, motivation and cognition. In W. K. Estes (Ed.), *Handbook of learning and cognitive processes.* Vol. 1. Hillsdale, N. J.: Lawrence Erlbaum Assoc., 1975. (a)

Bolles, R. C. *Learning theory.* New York: Holt, Rinehart & Winston, 1975. (b)

Bolles, R. C., Riley, A. L., & Laskowski, B. A further study of the learned safety effect in food-aversion learning. *Bulletin of the Psychonomic Society,* 1973, **1,** 190–192.

Borkhuis, M. L., Davis, R. T., & Medin, D. L. Confusion errors in monkey short term memory. *Journal of Comparative & Physiological Psychology,* 1971, **77,** 206–211.

Bower, G. H. A multicomponent theory of the memory trace. In K. W. Spence & J. T. Spence (Eds.), *The psychology of learning and motivation: Advances in research and theory.* Vol. 1. New York: Academic Press, 1967.

Bower, G. H. Stimulus-sampling theory of encoding variability. In A. W. Melton & E. Martin (Eds.), *Coding processes in human memory.* Washington, D.C.: Winston, 1972.

Bower, T. G. R. Object in the world of the infant. *Scientific American,* 1971, **225,** 30–38.

Brown, J. A. Some tests of the decay theory of immediate memory. *Quarterly Journal of Experimental Psychology,* 1958, **10,** 12–21.

Bryne, W.L. *Molecular approaches to learning and memory.* New York: Academic Press, 1970.

Bullock, M. A., & Richards, R. W. Disruption of fixed-ratio performance as a function of reinforcement delay. *Bulletin of the Psychonomic Society,* 1973, **1,** 49–52.

Burr, D. E. S., & Thomas, D. R. Effect of proactive inhibition upon postdiscrimination generalization gradient. *Journal of Comparative & Physiological Psychology,* 1972, **81,** 441–448.

Campbell, B. A., & Jaynes, J. Reinstatement. *Psychological Review,* 1966, **73,** 478–480.

Campbell, B. A., & Spear, N. E. Ontogeny of memory. *Psychological Review,* 1972, **79,** 215–236.

Capaldi, E. J. A sequential hypothesis of instrumental learning. In K. W. Spence & J. T. Spence (Eds.), *The psychology of learning and motivation.* Vol. 1. New York: Academic Press, 1967.

Capaldi, E. J. Memory and learning: A sequential viewpoint. In W. K. Honig & P. H. R. James (Eds.), *Animal memory.* New York: Academic Press, 1971.

Capaldi, E. J., & Cogan, D. Magnitude of reward and differential stimulus consequences. *Psychological Reports,* 1963, **13,** 85–86.

Capaldi, E. J., & Lynch, D. Repeated shifts in reward magnitude: Evidence in favor of an associational and absolute (noncontextual) interpretation. *Journal of Experimental Psychology,* 1967, **75,** 226–235.

Capaldi, E. J., & Spivey, J. E. Intertrial reinforcement and aftereffects at 24-hour intervals. *Psychonomic Science,* 1964, **1,** 181–182.

Carr, A. M. More studies with the white rat. I. Normal animals. *Journal of Animal Behavior,* 1917, **1,** 259–275.

Chiszar, D. A., & Spear, N. E. Stimulus change, reversal learning, and retention in the rat. *Journal of Comparative & Physiological Psychology,* 1969, **69,** 190–195.

Cogan, D., & Capaldi, E. J. Relative effects of delayed reinforcement and partial reinforcement on acquisition and extinction. *Psychological Reports,* 1961, **9,** 7–13.

Cowles, J. T. Food versus no food on the pre-delay trial of delayed response. *Journal of Comparative Psychology,* 1941, **32,** 153–164.

Crespi, L. P. Quantitative variation of incentive and performance in the white rat. *American Journal of Psychology,* 1942, **55,** 467–517.

Crum, J., Brown, W. L., & Bitterman, M. E. The effect of partial and delayed reinforcement on resistance to extinction. *American Journal of Psychology,* 1951, **64,** 228–237.

D'Amato, M. R. Delayed matching and short-term memory in monkeys. In G. H. Bower (Ed.), *The psychology of learning and motivation: Advances in research and theory.* Vol. 7. New York: Academic Press, 1973.

D'Amato, M. R., & Worsham, R. W. Delayed matching in the capuchin monkey with brief sample durations. *Learning & Motivation,* 1972, **3,** 304–312.

D'Amato, M. R., & Worsham, R. W. Retrieval cues and short-term memory in capuchin monkeys. *Journal of Comparative & Physiological Psychology,* 1974, **86,** 274–282.

Davenport, J. The interaction of magnitude and delay of reinforcement in spatial discrimination. *Journal of Comparative & Physiological Psychology,* 1962, **55,** 267–273.

Davis, R. T. Monkeys as perceivers. In L. A. Rosenblum (Ed.), *Primate behavior: Developments in field and laboratory research.* Vol. 3. New York: Academic Press, 1974. Pp. 1–263.

Davis, R. T., & Ruggiero, F. T. Memory in monkeys as a function of preparatory interval and pattern complexity of matrix displays. *American Journal of Physical Anthropology,* 1973, **38,** 573–578.

Deutsch, J. A. The physiological basis of memory. *Annual Review of Psychology,* 1969, **20,** 85–104.

Deustch, J. A. *The physiological basis of memory.* New York: Academic Press, 1973.

Dickerson, J. E., & Ellis, N. R. Effects of post-response stimulus duration upon discrimination learning in human subj^cts. *Journal of Experimental Psychology,* 1965, **69,** 528–583.

Domjan, M., & Bowman, T. G. Learned safety and the CS–US delay gradient in taste-aversion learning. *Learning & Motivation,* 1974, **5,** 409–423.

Domjan, M., & Wilson, N. E. Contribution of ingestive behaviors to fast aversion learning. *Journal of Comparative and Physiological Psychology,* 1972, **80,** 403–412. (a)

Domjan, M., & Wilson, N. E. Specificity of cue to consequence in aversion learning in the rat. *Psychonomic Science,* 1972, **26,** 143–145. (b)

Donegan, W. H., Whitlow, J. W., & Wagner, A. R. Posttrial reinstatement of the CS in Pavlovian conditioning: Facilitation or impairment of acquisition as a function of individual differences in responsiveness to the CS. Unpublished manuscript, 1975.

Dweck, C. S., & Wagner, A. R. Situational cues and correlation between CS and US as determinants of the conditioned emotional response. *Psychonomic Science,* 1970, **18,** 145–147.

Ehrenfreund, D. Generalization of secondary reinforcement in discrimination learning. *Journal of Comparative & Physiological Psychology,* 1954, **47,** 311–314.

Eibl-Eibesfeldt, I. *Ethology: The biology of behavior.* New York: Holt, Rinehart & Winston, 1970.

Eisenson, J. *Adult aphasia: assessment and treatment.* Englewood Cliffs, N. J.: Prentice-Hall, 1973.

Eninger, N. U. The role of generalized approach and avoidance tendencies in brightness discrimination. *Journal of Comparative and Physiological Psychology,* 1953, **69,** 528–533.

Estes, W. K. Statistical theory of distributional phenomena in learning. *Psychological Review,* 1955, **62,** 369–377. (a)

Estes, W. K. Statistical theory of spontaneous recovery and regression. *Psychological Review,* 1955, **62,** 145–154. (b)

Estes, W. K. An associative basis for coding and organization in memory. In A. W. Melton & E. Martin (Eds.), *Coding processes in human memory.* Washington, D.C.: Winston, 1972.

Estes, W. K. Memory and conditioning. In F. J. McGuigan & D. B. Lunsden (Eds.), *Contemporary approaches to conditioning and learning.* New York: Wiley, 1973.

Estes, W. K., & Straughan, J. H. Analysis of a verbal conditioning situation in terms of statistical learning theory. *Journal of Experimental Psychology,* 1954, **47**, 225–234.

Feigley, D. A., & Spear, N. E. Effect of age and punishment condition on long-term retention by the rat of active and passive avoidance learning. *Journal of Comparative and Physiological Psychology,* 1970, **73**, 515–526.

Ferster, C. B., & Hammer, C. Variables determining the effects of delay in reinforcement. *Journal of the Experimental Analysis of Behavior,* 1965, **8**, 243–254.

Fitzgerald, R. D., & Davis, R. T. The role of preference and reward in the selection of discriminanda by naive and sophisticated rhesus monkeys. *Journal of Genetic Psychology,* 1960, **97**, 227–235.

Flagg, S. F. *Transformation of position and short-term memory in rhesus monkeys.* Unpublished doctoral dissertation, Washington State University, Pullman, 1975.

Fletcher, H. J. The delayed response problem. In A. M. Schrier, H. F. Harlow, & F. Stollnitz (Eds.), *Behavior of nonhuman primates: Modern research trends.* Vol. 1. New York: Academic Press, 1965.

Frankmann, J. P. Effect of amount of interpolated learning and time interval before test on retention in rats. *Journal of Experimental Psychology,* 1957, **54**, 462–466.

French, G. M. Associative problems. In A. M. Schrier, H. F. Harlow, & F. Stollnitz (Eds.), *Behavior of nonhuman primates: Modern research trends.* Vol. 1. New York: Academic Press, 1965.

Furth, H. G. *Thinking without language: Psychological implications of deafness.* New York: Free Press, 1966.

Garcia, J., & Koelling, R. A. Relation of cue to consequence in avoidance learning. *Psychonomic Science,* 1966, **4**, 123–124.

Garcia, J., McGowan, B. K., Ervin, F. R., & Koelling, R. A. Cues: Their relative effectiveness as a function of the reinforcer. *Science,* 1968, **160**, 794–795.

Gardiner, J. M., Craik, F. I. M., & Birtwistle, J. Retrieval cues and release from proactive inhibition. *Journal of Verbal Learning & Verbal Behavior,* 1972, **11**, 778–783.

Gibbon, J., Berryman, R., & Thompson, R. L. Contingencies, spaces and measures in classical and instrumental conditioning. *Journal of the Experimental Analysis of Behavior,* 1974, **21**, 585–605.

Girden, E. Conditioning and problem-solving behavior. *American Journal of Psychology,* 1938, **51**, 677–687.

Gleitman, H. Forgetting of long-term memories in animals. In W. K. Honig & P. H. R. James (Eds.), *Animal memory.* New York: Academic Press, 1971.

Gleitman, H. Getting animals to understand the experimenter's instructions. *Animal Learning & Behavior,* 1974, **2**, 1–5.

Gleitman, H., & Jung, L. Retention in rats: The effect of proactive interference. *Science,* 1963, **142**, 1683–1684.

Gleitman, H., & Steinman, F. Depression effect as a function of retention interval before and after shift in reward magnitude. *Journal of Comparative and Physiological Psychology,* 1964, **57**, 158–160.

Gleitman, H., Wilson, W. A., Jr., Herman, M. M., & Rescorla, R. A. Massing and within-delay position as factors in delayed-response performance. *Journal of Comparative & Physiological Psychology,* 1963, **56**, 445–451.

Glendenning, R. L., & Meyer, D. R. Motivationally related retroactive interference in discrimination learning by rats. *Journal of Comparative & Physiological Psychology,* 1971, **75**, 153–156.

Grant, D. S. Effect of presentation time on long-delay matching in the pigeon. Unpublished manuscript, 1975. (a)

Grant, D. S. Proactive inhibition in pigeon short-term memory. *Journal of Experimental Psychology: Animal Behavior Processes,* 1975, **104,** 207–220. (b)

Grant, D. S. Proactive interference and interfering memory decay time. Unpublished manuscript, 1975. (c)

Grant, D. S., & Roberts, W. A. Trace interaction in pigeon short-term memory. *Journal of Experimental Psychology,* 1973, **101,** 21–29.

Gray, P. Effects of adrenocorticotropic hormone on conditioned avoidance in rats interpreted as state-dependent learning. *Journal of Comparative & Physiological Psychology,* 1975, **88,** 281–284.

Gray, T., & Appignanesi, A. A. Compound conditioning: Elimination of the blocking effect. *Learning & Motivation,* 1973, **4,** 374–380.

Greenspoon, J., & Ranyard, R. Stimulus conditions and retroactive inhibition. *Journal of Experimental Psychology,* 1957, **53,** 55–59.

Grice, G. R. The relation of secondary reinforcement to delayed reward in visual discrimination learning. *Journal of Experimental Psychology,* 1948, **38,** 1–16.

Guthrie, E. R. *The psychology of learning.* New York: Harper, 1935.

Guthrie, E. R. Association and the law of effect. *Psychological Review,* 1940, **47,** 127–148.

Harlow, H. F. The formation of learning sets. *Psychological Review,* 1949, **56,** 51–56.

Harlow, H. F., Harlow, M. K., Rueping, R. R., & Mason, W. A. Performance of infant monkeys on discrimination learning, delayed response, and discrimination learning set. *Journal of Comparative & Physiological Psychology,* 1960, **53,** 113–121.

Harlow, H. F., & Hicks, L. H. Discrimination learning theory: Uniprocess vs. duoprocess. *Psychological Review,* 1957, **64,** 104–109.

Harlow, H. F., Uehling, H., & Maslow, A. H. Comparative behavior of primates. I. Delayed reaction tests on primates from the lemur to the orang-outan. *Journal of Comparative Psychology,* 1932, **13,** 313–343.

Harlow, H. F., & Warren, J. M. Formation and transfer of discrimination learning sets. *Journal of Comparative & Physiological Psychology,* 1952, **45,** 482–489.

Harris, A. V. *Visual masking in monkeys.* Unpublished doctoral dissertation, University of South Dakota, Vermillion, 1970.

Hawkins, H. L., Pardo, V. J., & Cox, R. D. Proactive interference in short-term recognition: Trace interaction or competition? *Journal of Experimental Psychology,* 1972, **92,** 43–48.

Hearst, E., & Peterson, G. B. Transfer of conditioned excitation and inhibition from one operant response to another. *Journal of Experimental Psychology,* 1973, **99,** 360–368.

Hecaen, H. Clinoco-anatomical and neurolinguistic aspects of aphasia. In G. A. Talland & N. C. Waugh (Eds.), *The pathology of memory.* New York: Academic Press, 1969.

Heider, F. *The psychology of interpersonal relations.* New York: Wiley, 1958.

Hellyer, S. Frequency of stimulus presentation and short-term decrement in recall. *Journal of Experimental Psychology,* 1962, **64,** 650.

Heth, C. D., & Rescorla, R. A. Simultaneous and backward fear conditioning in the rat. *Journal of Comparative & Physiological Psychology,* 1973, **82,** 434–443.

Herman, L. M. Interference and auditory short-term memory in the bottlenosed dolphin. *Animal Learning & Behavior,* 1975, **3,** 43–48.

Herman, L. M., & Gordon, J. A. Auditory delayed matching in the bottlenose dolphin. *Journal of the Experimental Analysis of Behavior,* 1974, **21,** 19–26.

Herrnstein, R. J. On the law of effect. *Journal of the Experimental Analysis of Behavior,* 1970, **13,** 243–266.

Hertz, M. Figural perception in the jay bird. In W. D. Ellis (Edited and English translation), *A source book of Gestalt psychology.* New York: Humanities Press, 1955.

Hoffman, H. S. Stimulus factors in conditioned suppression. In B. A. Campbell & R. M. Church (Eds.), *Punishment and aversive behavior.* New York: Appleton-Century-Crofts, 1969.

Holloway, F. A., & Wansley, R. Multiphasic retention deficits at periodic intervals after passive-avoidance learning. *Science,* 1973, **180,** 208–210.

Honig, W. K., & James, P. H. R. (Eds.), *Animal memory.* New York: Academic Press, 1971.

Howard, R. L., Glendenning, R. L., & Meyer, D. R. Motivational control of retrograde amnesia: Further explorations and effects. *Journal of Comparative & Physiological Psychology,* 1974, **86,** 187–192.

Hull, C. L. *Principles of behavior.* New York: Appleton Century, 1943.

Hull, C. L. *A behavior system.* New Haven, Conn.: Yale University Press, 1952.

Hunter, W. S. The delayed reaction in animals and children. *Behavior Monographs,* 1913, **2,** No. 1, 1–86.

Hunter, W. S. The sensory control of the maze habit in the white rat. *Journal of Genetic Psychology,* 1929, **36,** 505–537.

Jahnke, J. C. Restrictions on the Ranschburg effect. *Journal of Experimental Psychology,* 1974, **103,** 183–185.

Jarrard, L. E. (Ed.) *Cognitive processes of nonhuman primates.* New York: Academic Press, 1971.

Jarrard, L. E., & Moise, S. L. Short-term memory in the stumptail macaque: Effect of physical restraint of behavior on performance. *Learning & Motivation,* 1970, **1,** 267–275.

Jarrard, L. E., & Moise, S. L. Short-term memory in the monkey. In L. E. Jarrard (Ed.), *Cognitive processes of nonhuman primates.* New York: Academic Press, 1971.

Jarvik, M. E. Effects of chemical and physical treatments on learning and memory. *Annual Review of Psychology,* 1972, **23,** 457–486.

Jarvik, M. E., Goldfarb, T. L., & Carley, J. L. Influence of interference on delayed matching in monkeys. *Journal of Experimental Psychology,* 1969, **81,** 1–6.

Kalat, J. W., & Rozin, P. "Learned safety" as a mechanism in long-delay taste-aversion learning in rats. *Journal of Comparative & Physiological Psychology,* 1973, **83,** 198–207.

Kamin, L. J. The retention of an incomplete learning avoidance response. *Journal of Comparative & Physiological Psychology,* 1957, **50,** 457–460.

Kamin, L. J. Backward conditioning and the conditioned emotional response. *Journal of Comparative & Physiological Psychology,* 1963, **56,** 517–519.

Kamin, L. J. Attention-like processes in classical conditioning. In M. R. Jones (Ed.), *Miami symposium on the prediction of behavior: Aversive stimulation.* Miami: University of Miami Press, 1968.

Kamin, L. J. Predictability, surprise, attention, and conditioning. In B. A. Campbell & R. M. Church (Eds.), *Punishment and aversive behavior.* New York: Appleton-Century-Crofts, 1969.

Keppel, G. Forgetting. In C. P. Duncan, L. Sechrest, & A. W. Melton (Eds.), *Human memory: Festschrift in honor of Benton J. Underwood.* New York: Appleton-Century-Crofts, 1972.

Keppel, G., & Underwood, B. J. Proactive inhibition in short-term retention of single items. *Journal of Verbal Learning & Verbal Behavior,* 1962, **1,** 153–161.

Kimble, G. A. *Hilgard and Marquis' conditioning and learning.* New York: Appleton-Century-Crofts, 1961.

Klein, S. B., & Spear, N. E. Influence of age on short-term retention of active avoidance learning in rats. *Journal of Comparative & Physiological Psychology,* 1969, **69,** 583–589.

Knapp, R. K., Kause, R. H., & Perkins, C. C. Immediate versus delayed shock in T-maze performance. *Journal of Experimental Psychology,* 1959, **58,** 357–362.

Konorski, J. One method of physiological investigation of recent memory in animals. *Bulletin de l'acadanne Polinise des Science,* 1959, **7,** 115.

Koppenaal, R. J., & Jagoda, E. Proactive inhibition of a maze position habit. *Journal of Experimental Psychology,* 1968, **76,** 664–668.

Krane, R. V., & Wagner, A. R. Taste aversion learning with a delayed shock US: Implications for the "Generality of the Laws of Learning." *Journal of Comparative & Physiological Psychology,* 1975, **88,** 882–889.

Krechevsky, I. "Hypotheses" in rats. *Psychological Review,* 1932, **39,** 516–532.

Kroll, N. E. A., Parks, T., Parkinson, S. R., Bieber, S. L., & Johnson, H. L. Short-term memory while shadowing: Recall of visually and aurally presented letters. *Journal of Experimental Psychology,* 1970, **85,** 220–224.

Ladieu, G. The effect of length of the delay interval upon delayed alternation in the albino rat. *Journal of Comparative Psychology,* 1944, **37,** 273–286.

Lashley, K. S. Conditional reactions in the rat. *Journal of Psychology,* 1938, **6,** 311–324.

Lawrence, D. H., & Hommel, L. The influence of differential goal boxes on discrimination learning involving delay of reinforcement. *Journal of Comparative & Physiological Psychology,* 1961, **54,** 552–555.

Leary, R. W. The rewarded, the unrewarded, the chosen, and the unchosen. *Psychological Reports,* 1956, **2,** 91–97.

Leary, R. W. The temporal factor in reward and nonreward of monkeys. *Journal of Experimental Psychology,* 1958, **56,** 294–296. (a)

Leary, R. W. Analysis of serial-discrimination learning by monkeys. *Journal of Comparative & Physiological Psychology,* 1958, **51,** 82–86. (b)

Lett, B. T. Delayed reward learning: Disproof of the traditional theory. *Learning & Motivation,* 1973, **4,** 237–246.

Lett, B. T. Visual discrimination learning with a 1-min delay of reward. *Learning & Motivation,* 1974, **5,** 174–181.

Lett, B. T. Long delay learning in the T-maze. *Learning & Motivation,* 1975, **6,** 80–90.

LeVere, T. E. Linear pattern completion by chimpanzees. *Psychonomic Science,* 1966, **5,** 15–16.

LeVere, T. E. The effects of pre- and post-response stimulus sampling on discrimination performance. In D. Stark, R. Schneider, & H. J. Kuhn (Eds.), *Progress in primatology.* Stuttgart, Germany: Gustav Fisher Verlag, 1967.

LeVere, T. E. The failure of stimulus-reinforcement spatial discontiguity to influence performance under optimal conditions of stimulus sampling. *Psychonomic Science,* 1968, **11,** 179–180. (a)

LeVere, T. E. Cue size and learning set performance with S–R induced sampling biases eliminated. *Journal of Comparative & Physiological Psychology,* 1968, **65,** 362–365. (b)

LeVere, T. E., & Bartus, R. T. APDA: A discontiguous S–R automated primate discrimination apparatus. *Behavior Research Methodology & Instrumentation,* 1969, **1,** 259–262.

LeVere, T. E., & Bartus, R. T. Stimulus information and primate discrimination learning: Pre-response utilization of stimulus information. *Journal of Comparative & Physiological Psychology,* 1971, **77,** 200–205.

LeVere, T. E., & Bartus, R. T. Stimulus information and primate discrimination learning: Utilization of pre-response stimulus information following acquisition. *Journal of Comparative & Physiological Psychology,* 1972, **79,** 432–437.

LeVere, T. E., & Bartus, R. T. Stimulus information and primate discrimination learning: The influence of post-response stimulus information. *Learning & Motivation,* 1973, **4,** 305–313.

Levine, S. The role of irrelevant drive stimuli in learning. *Journal of Experimental Psychology,* 1953, **45**, 410–416.

Lewis, D. J., & Bregman, N. J. The source of the cues for cue-dependent amnesia. *Journal of Comparative & Physiological Psychology,* 1973, **85**, 421–426.

Lewis, D. J., Bregman, N. J., & Mahan, J. J. Cue-dependent amnesia in rats. *Journal of Comparative & Physiological Psychology,* 1972, **81**, 243–247.

Liddell, H. S., James, W. T., & Anderson, O. P. The comparative physiology of the conditioned motor reflex based on experiments with the pig, dog, sheep, goat and rabbit. *Comparative Psychology Monographs,* 1934, **11**, (1).

Loess, H. Proactive inhibition in short-term memory. *Journal of Verbal Learning & Verbal Behavior,* 1964, **3**, 362–368.

Loess, H., & Waugh, N. C. Short-term memory and intertrial interval. *Journal of Verbal Learning & Verbal Behavior,* 1967, **6**, 455–460.

Logan, F. A. *Incentive.* New Haven, Conn.: Yale University Press, 1960.

Loucks, R. B. Efficacy of the rats motor cortex in delayed alternation. *Journal of Comparative Neurology,* 1931, **53**, 511–567.

Mackintosh, N. J. *The psychology of animal learning.* New York: Academic Press, 1974.

Madigan, S. W. Intraserial repetition and coding processes in free recall. *Journal of Verbal Learning & Verbal Behavior,* 1969, **8**, 828–835.

Maier, N. R. F. Reasoning and learning. *Psychological Review,* 1931, **38**, 332–346.

Maier, N. R. F., & Schneirla, T. C. *Principles of animal psychology.* New York: McGraw-Hill, 1935.

Martin, E. Stimulus meaningfulness and paired-associate transfer: An encoding variability hypothesis. *Psychological Review,* 1968, **75**, 421–441.

Martin, I., & Levey, A. B. *The genesis of the classical conditioned response.* Oxford: Pergamon Press, 1969.

Mason, M., & Wilson, M. Temporal differentiation and recognition memory for visual stimuli in rhesus monkeys. *Journal of Experimental Psychology,* 1974, **103**, 383–390.

Mayer, M. J., & Ross, L. E. The effects of stimulus complexity, interstimulus interval, and masking conditions in differential eyelid conditioning. *Journal of Experimental Psychology,* 1969, **81**, 469–474.

McAllister, D. E., McAllister, W. R., Brooks, C. I., & Goldman, J. A. Magnitude and shift of reward in instrumental aversive learning in rats. *Journal of Comparative & Physiological Psychology,* 1972, **80**, 490–501.

McAllister, W. R., & McAllister, D. E. Behavioral measurement of conditioned fear. In F. R. Brush (Ed.), *Aversive conditioning and learning.* New York: Academic Press, 1971.

McGaugh, J. L., & Herz, M. J. *Memory consolidation.* San Francisco: Albion, 1972.

McGeoch, J. A. Forgetting and the law of disuse. *Psychological Review,* 1932, **39**, 352–370.

McGeoch, J. A. *The psychology of human learning: an introduction.* New York: Longmans, Green, 1942.

Medin, D. L. Form perception and pattern reproduction in monkeys. *Journal of Comparative & Physiological Psychology,* 1969, **68**, 412–419.

Medin, D. L. Role of reinforcement in discrimination learning set in monkeys. *Psychological Bulletin,* 1972, **77**, 305–318. (a)

Medin, D. L. Evidence for short- and long-term memory in monkeys. *American Journal of Psychology,* 1972, **85**, 117–119. (b)

Medin, D. L. The comparative study of memory. *Journal of Human Evolution,* 1974, **3**, 455–463.

Medin, D. L. A theory of context in discrimination learning. In G. H. Bower (Ed.), *The psychology of learning and motivation.* Vol. 9. New York: Academic Press, 1975. Pp. 263–314.

Medin, D. L., & Cole, M. Comparative psychology and human cognition. In W. K. Estes (Ed.), *Handbook of learning and cognitive processes.* Hillsdale, N. J.: Lawrence Erlbaum Assoc., 1975. Pp. 111–149.

Medin, D. L., & Davis, R. T. Memory. In A. M. Schrier & F. Stollnitz (Eds.), *Behavior of nonhuman primates: Modern research trends.* Vol. 5. New York: Academic Press, 1974.

Mello, N. Alcohol effects on delayed matching to sample performance by rhesus monkeys. *Physiology & Behavior,* 1971, **7**, 77–101.

Melton, A. W. The situation with respect to the spacing of repetitions and memory. *Journal of Verbal Learning & Verbal Behavior,* 1970, **9**, 596–606.

Melton, A. W., & Martin, E. (Eds.), *Coding processes in human memory.* Washington, D.C.: Winston, 1972.

Meyer, D. R. Access to engrams. *American Psychologist,* 1972, **27**, 124–131.

Meyer, D. R., Treichler, F. R., & Meyer, P. M. Discrete-trial training techniques and stimulus variables. In A. M. Schrier, H. F. Harlow, & F. Stollnitz (Eds.), *Behavior of nonhuman primates: Modern research trends.* Vol. 1. New York: Academic Press, 1965.

Miles, R. C. Species differences in "transmitting" spatial location information. In L. E. Jarrard (Ed.), *Cognitive processes of nonhuman primates.* New York: Academic Press, 1971.

Miller, R. R., & Springer, A. D. Amnesia, consolidation and retrieval. *Psychological Review,* 1973, **80**, 69–79.

Mischel, W. Processes in delay of gratification. In L. Berkowitz (Ed.), *Advances in experimental social psychology.* Vol. 7. New York: Academic Press, 1974.

Mishkin, M., & Delacour, J. An analysis of short-term visual memory in the monkey. *Journal of Experimental Psychology: Animal Behavior Processes,* 1975, **1**, 326–334.

Mishkin, M., & Weiskrantz, L. Effects of delaying reward on visual-discrimination performance in monkeys with frontal lesions. *Journal of Comparative & Physiological Psychology,* 1958, **51**, 276–281.

Mitchell, D., Scott, D. W., & Williams, K. D. Container neophobia and the rat's preference for earned food. *Behavioral Biology,* 1973, **9**, 613–624.

Moise, S. L. Short-term retention in *Maccaca speciosa* following interpolated activity during matching from sample. *Journal of Comparative & Physiological Psychology,* 1970, **73**, 506–514.

Morse, P. A., & Snowdon, C. T. An investigation of categorical speech discrimination by rhesus monkeys. *Perception & Psychophysics,* 1975, **17**, 9–16.

Moscovitch, A., & Lolordo, V. M. Role of safety in the Pavlovian backward fear conditioning procedure. *Journal of Comparative & Physiological Psychology,* 1968, **66**, 673–678.

Moss, E. M., & Harlow, H. F. The role of reward in discrimination learning in monkeys. *Journal of Comparative & Physiological Psychology,* 1947, **40**, 333–342.

Motiff, J. P. *Learning and retention of redundant patterns by monkeys.* Unpublished doctoral dissertation, University of South Dakota, Vermillion, 1969.

Motiff, J. P., DeKock, A. R., & Davis, R. T. Concealment of stimuli during delay in the delayed-response problem. *Perceptual & Motor Skills,* 1969, **29**, 788–790.

Mowrer, O. H., & Aiken, E. G. Contiguity vs. drive-reduction in conditioned fear: Temporal variations in conditioned and unconditioned stimulus. *American Journal of Psychology,* 1954, **67**, 26–38.

Munn, N. L. *Handbook of psychological research on the rat.* Boston: Houghton Mifflin, 1950.

Murdock, B. B. The immediate retention of unrelated words. *Journal of Experimental Psychology,* 1960, **60**, 222–234.

Murdock, B. B. *Human memory: Theory and data.* Hillsdale, N. J.: Lawrence Erlbaum Assoc., 1974.

Nachman, M., & Jones, D. R. Learned taste aversions over long delays in rats: The role of learned safety. *Journal of Comparative & Physiological Psychology,* 1974, **86,** 949–956.

Neimark, E., & Estes, W. K. *Stimulus sampling theory.* San Francisco: Holden Day, 1967.

Nissen, H. W., Carpenter, C. R., & Cowles, J. T. Stimulus versus response-differentiation in delayed reactions of chimpanzees. *Journal of Genetic Psychology,* 1936, **48,** 112–136.

Norman, D. A. (Ed.) *Models of human memory.* New York: Academic Press, 1970.

Oppenheimer, J. R., & Lang, G. E. Cebus monkeys: Effect on branching of Gustavia trees. *Science,* 1969, **165,** 187–188.

Orne, M. T. The mechanisms of hypnotic age regression: An experimental study. *Journal of Abnormal & Social Psychology,* 1951, **46,** 213–225.

Oscar-Berman, M., Heywood, S. P., & Gross, C. G. Eye orientation during visual discrimination learning by monkeys. *Neuropsychologia,* 1971, **9,** 351–358.

Osgood, C. E. *Method and theory in experimental psychology.* New York: Oxford University Press, 1953.

Ost, J. W. P. Consolidation disruption and inhibition in classical conditioning. *Psychological Review,* 1969, **72,** 379–383.

Overmier, J. B., & Bull, J. A. Influences of appetitive Pavlovian conditioning upon avoidance behavior. In J. H. Reynierse (Ed.), *Current issues in animal learning.* Lincoln: University of Nebraska Press, 1970.

Overton, D. A. State-dependent learning produced by depressant and atropine-like drugs. *Psychopharmacologia,* 1966, **10,** 6–31.

Overton, D. A. State-dependent learning produced by alcohol and its relevance to alcoholism. In B. Kissin & H. Begleiter (Eds.), *The biology of alcoholism.* Vol. 2. *Physiology and behavior.* New York: Plenum Press, 1972.

Paivio, A., & Bleasdale, F. Visual short-term memory: A methodological caveat. *Canadian Journal of Psychology,* 1974, **28,** 24–31.

Pavlov, I. P. *Conditioned reflexes.* (Translated by G. V. Anrep.) London: Oxford University Press, 1927.

Pavlov, I. P. *Lectures on conditioned reflexes.* New York: International Publisher, 1928.

Peterson, L. R., & Peterson, M. J. Short-term retention of individual verbal items. *Journal of Experimental Psychology,* 1959, **58,** 193–198.

Petrinovich, L., & Bolles, R. C. Delayed alternation: Evidence for symbolic processes in the rat. *Journal of Comparative & Physiological Psychology,* 1957, **50,** 363–365.

Piaget, J., & Inhelder, B. *Memory and intelligence.* New York: Basic Books, 1973.

Polidora, V. J., & Fletcher, H. J. An analysis of the importance of S–R spatial contiguity for proficient primate discrimination performance. *Journal of Comparative & Physiological Psychology,* 1964, **57,** 224–230.

Pribram, K. H. *Languages of the brain: Experimental paradoxes and principles in neuropsychology.* Englewood Cliffs, N. J.: Prentice-Hall, 1971.

Pribram, K. H., & Broadbent, D. E. *Biology of memory.* New York: Academic Press, 1970.

Pribram, K. H., & Mishkin, M. Analysis of the effects of frontal lesions in monkey: III. Object alternation. *Journal of Comparative & Physiological Psychology,* 1956, **49,** 41–45.

Pschirrer, M. E. Goal events as discriminative stimuli over extended intertrial intervals. *Journal of Experimental Psychology,* 1972, **96,** 425–432.

Rand, G., & Wapner, S. Postural status as a factor in memory. *Journal of Verbal Learning & Verbal Behavior,* 1967, **6,** 268–271.

Razran, G. Backward conditioning. *Psychological Bulletin,* 1956, **53,** 55–69.

Reberg, D. Compound tests for excitation in early acquisition and after prolonged extinction of conditioned suppression. *Learning & Motivation,* 1972, **3,** 246–258.

Renner, K. E. Temporal integration: An incentive approach to conflict resolution. In B. Maher (Ed.), *Progress in experimental personality research.* Vol. 4. New York: Academic Press, 1967.

Renner, K. E. Temporal integration: Amount of reward and relative utility of immediate and delayed outcomes. *Journal of Comparative & Physiological Psychology,* 1968, **65,** 182–186.

Rescorla, R. A. Informational variables in Pavlovian conditioning. In G. H. Bower (Ed.), *The psychology of learning and motivation.* Vol. 6. New York: Academic Press, 1972.

Rescorla, R. A., & Heth, C. D. Reinstatement of fear to an extinguished conditioned stimulus. *Journal of Experimental Psychology: Animal Behavior Processes,* 1975, **104,** 88–96.

Rescorla, R. A., & Wagner, A. R. A theory of Pavlovian conditioning: Variations in the effectiveness of reinforcement and nonreinforcement. In A. H. Black & W. F. Prokasy (Eds.), *Classical conditioning II.* New York: Appleton-Century-Crofts, 1972.

Restle, F. *Psychology of judgment and choice.* New York: Wiley, 1961.

Revusky, S. The role of interference in association over a delay. In W. K. Honig & P. H. R. James (Eds.), *Animal memory.* New York: Academic Press, 1971.

Revusky, S., & Bedarf, E. W. Association of illness with prior ingestion of novel foods. *Science,* 1967, **155,** 219–220.

Revusky, S., & Garcia, J. Learned associations over long delays. In G. Bower & J. T. Spence (Eds.), *Psychology of learning and motivation.* Vol. 4. New York: Academic Press, 1970.

Riley, A. L. The relative contribution of neophobia to conditioned taste aversions. Unpublished doctoral dissertation, University of Washington, Seattle, 1974.

Riopelle, A. J. Performance of rhesus monkeys on spatial delayed response (indirect method). *Journal of Comparative & Physiological Psychology,* 1959, **52,** 746–753.

Riopelle, A. J., & Chinn, R. McC. Position habits and discrimination learning by monkeys. *Journal of Comparative & Physiological Psychology,* 1961, **54,** 178–180.

Riopelle, A. J., & Churukian, G. A. The effect of varying the intertrial interval in discrimination learning by normal and brain-operated monkeys. *Journal of Comparative & Physiological Psychology,* 1958, **51,** 119–125.

Riopelle, A. J., & Copelan, E. L. Discrimination reversal to a sign. *Journal of Experimental Psychology,* 1954, **48,** 143–145.

Robbins, D., & Bush, C. T. Memory in great apes. *Journal of Experimental Psychology,* 1973, **97,** 344–348.

Roberts, W. A. *A review of findings on the delayed-response problem as they pertain to the problem of short-term memory in animals.* (Research Bulletin No. 228). Unpublished manuscript, University of Western Ontario, London, 1972. (a)

Roberts, W. A. Short-term memory in the pigeon: Effects of repetition and spacing. *Journal of Experimental Psychology,* 1972, **94,** 74–83. (b)

Roberts, W. A. Spatial separation and visual differentiation of cues of factors influencing short-term memory in the rat. *Journal of Comparative & Physiological Psychology,* 1972, **78,** 284–291. (c)

Roberts, W. A. Free recall of word lists varying in length and rate of presentation: A test of total-time hypotheses. *Journal of Experimental Psychology,* 1972, **92,** 365–372. (d)

Roberts, W. A. Spaced repetition facilitates short-term retention in the rat. *Journal of Comparative & Physiological Psychology,* 1974, **86,** 164–171.

Roberts, W. A. A failure to replicate visual discrimination learning with a 1-min delay of reward. Unpublished manuscript, 1975.

Roberts, W. A., & Grant, D. S. Short-term memory in the pigeon with presentation time precisely controlled. *Learning & Motivation,* 1974, **5,** 393–408.

Roediger, H. L. Inhibiting effects of recall. *Memory & Cognition,* 1974, **2,** 261–269.

Rose, R. M., & Vitz, P. C. The role of runs in probability learning. *Journal of Experimental Psychology,* 1966, **72,** 751–760.

Rudel, R. G., & Teuber, H. L. Discrimination of line in children. *Journal of Comparative & Physiological Psychology,* 1963, **56,** 892–898.

Rudy, J. W. Stimulus selection in animal conditioning and paired-associate learning: Variations in the associative process. *Journal of Verbal Learning & Verbal Behavior,* 1974, **13,** 282–296.

Rundus, D. Analysis of rehearsal processes in free recall. *Journal of Experimental Psychology,* 1971, **89,** 63–77.

Rundus, D. Negative effects of using list items as recall cues. *Journal of Verbal Learning & Verbal Behavior,* 1973, **12,** 43–50.

Ruggiero, F. T. Coding processes and contextual cues in monkey short term memory. (Doctoral dissertation, Washington State University, 1974). *Dissertation Abstracts International,* 1974, **35B,** No. 1, 555–556. (University Microfilms No. 74-16, 392).

Schneck, M. R., & Warden, C. J. A comprehensive survey of the experimental literature on animal retention. *Pedagogical Seminary & Journal of Genetic Psychology,* 1929, **36,** 1–20.

Schneiderman, N. Response system divergencies in aversive classical conditioning. In A. H. Black & W. F. Prokasy (Eds.), *Classical conditioning II.* New York: Appleton-Century-Crofts, 1972.

Schrier, A. M. Effect of location of the conditional cue on conditional discrimination learning by monkeys (*Macaca mulatta*). *Learning & Motivation,* 1970, **1,** 207–217.

Schrier, A. M., & Wing, T. G. Eye movements of monkeys during brightness discrimination and discrimination reversal. *Animal Learning & Behavior,* 1973, **1,** 145–150.

Seligman, M. E. P. On the generality of the laws of learning. *Psychological Review,* 1970, **77,** 406–418.

Seward, J. P. An experimental analysis of latent learning. *Journal of Experimental Psychology,* 1949, **39,** 177–186.

Sheridan, C. L., Horel, J. A., & Meyer, D. R. Effects of response-induced stimulus change on primate discrimination learning. *Journal of Comparative & Physiological Psychology,* 1962, **55,** 511–514.

Shettleworth, S. J. Food reinforcement and the organization of behavior in golden hamsters. In R. A. Hinde & J. Stevenson-Hinde (Eds.), *Constraints on learning.* London: Academic Press, 1973.

Shiffrin, R. M. Forgetting: trace erosion or retrieval failure? *Science,* 1970, **168,** 1601–1603.

Shimp, C. P., & Moffitt, M. Short-term memory in the pigeon: Stimulus–response associations. *Journal of the Experimental Analysis of Behavior,* 1974, **22,** 507–512.

Sidman, M. Generalization gradients and stimulus control in delayed matching-to-sample. *Journal of the Experimental Analysis of Behavior,* 1969, **12,** 745–757.

Siegel, S., & Domjan, M. Backward conditioning as an inhibitory procedure. *Learning & Motivation,* 1971, **2,** 1–11.

Silvestri, R., Rohrbaugh, M. J., & Riccio, D. C. Conditions influencing the retention of learned fear in young rats. *Developmental Psychology,* 1970, **2,** 380–395.

Skinner, B. F. *The behavior of organisms.* New York: Century, 1938.

Smith, M. C., Coleman, S. R., & Gormezano, I. Classical conditioning of the rabbit's nictitating membrane response at backward, simultaneous, and forward CS–US intervals. *Journal of Comparative & Physiological Psychology,* 1969, **69,** 226–231.

Spear, N. E. Retention of reinforcer magnitude. *Psychological Review,* 1967, **74,** 216–234.

Spear, N. E. Forgetting as retrieval failure. In W. K. Honig & P. H. R. James (Eds.), *Animal memory.* New York: Academic Press, 1971.

Spear, N. E. Retrieval of memory in animals. *Psychological Review,* 1973, **80,** 163–194.

Spear, N. E. Retrieval of memories. In W. K. Estes (Ed.), *Handbook of learning and cognitive processes.* Vol. 4, *Memory processes.* Hillsdale, N. J.: Lawrence Erlbaum Assoc., 1976.

Spear, N. E., Gordon, W. C., & Chiszar, D. A. Interaction between memories in the rat: Effect of degree of prior conflicting learning on forgetting after short intervals. *Journal of Comparative & Physiological Psychology,* 1972, **78,** 471–477.

Spear, N. E., Gordon, W. C., & Martin, P. A. Warm-up decrement as failure in memory retrieval in the rat. *Journal of Comparative & Physiological Psychology,* 1973, **85,** 601–614.

Spear, N. E., Klein, S. B., & Riley, E. P. The Kamin effect as "state-dependent" learning: Memory retrieval failure in the rat. *Journal of Comparative & Physiological Psychology,* 1971, **74,** 416–425.

Spence, K. W. The nature of discrimination learning in animals. *Psychological Review,* 1936, **43,** 427–449.

Spence, K. W. The differential response in animals to stimuli varying within a single dimension. *Psychological Review,* 1937, **44,** 430–444.

Spence, K. W. The role of secondary reinforcement in delayed reward learning. *Psychological Review,* 1947, **54,** 1–8.

Spooner, A., & Kellogg, W. N. The backward conditioning curve. *American Journal of Psychology,* 1947, **60,** 321–334.

Standing, L., Conezio, J., & Haber, R. N. Perception and memory for pictures: Single trial learning of 2500 visual stimuli. *Psychonomic Science,* 1970, **19,** 73–74.

Stepien, L. S., & Cordeau, J. P. Memory in monkeys for compound stimuli. *American Journal of Psychology,* 1960, **73,** 388–395.

Stollnitz, F. Spatial variables, observing responses, and discrimination learning sets. *Psychological Review,* 1965, **72,** 247–261.

Stroebel, C. F. Behavioral aspects of circadian rhythms. In J. Zubin & H. F. Hunt (Eds.), *Comparative psychopathology.* New York: Grune & Stratton, 1967.

Sutherland, N. S. Visual discrimination of orientation by octopus. *British Journal of Psychology,* 1957, **48,** 55–71.

Sutherland, N. S., & Mackintosh, N. J. Mechanisms of animal discrimination learning. New York: Academic Press, 1971.

Switzer, C. A. Backward conditioning of the lid reflex. *Journal of Experimental Psychology,* 1930, **13,** 76–97.

Terry, W. S., & Wagner, A. R. Short-term memory for "surprising" vs. "expected" USs in Pavlovian conditioning. *Journal of Experimental Psychology: Animal Behavior Processes,* 1975, **104,** 122–133.

Testa, T. J. Causal relationships and the acquisition of avoidance responses. *Psychological Review,* 1974, **81,** 491–505.

Tinklepaugh, O. L. An experimental study of representative factors in monkeys. *Journal of Comparative Psychology,* 1928, **8,** 197–236.

Tinklepaugh, O. L. Multiple delayed reaction with chimpanzees and monkeys. *Journal of Comparative Psychology,* 1932, **13,** 207–243.

Tolman, E. C., & Honzik, C. H. Introduction and removal of reward and maze performance in rats. *University of California Publications in Psychology,* 1930, **4,** 257–275. (a)

Tolman, E. C., & Honzik, C. H. "Insight" in rats. *University of California Publications in Psychology,* 1930, **4,** 215–232. (b)

Tompkins, S. S. A theory of memory. In J. S. Antrobus (Ed.), *Cognition and affect.* Boston: Little, Brown, 1970. Pp. 59–130.

Treichler, F. R., Hann, B., & Way, S. J. Effects of response-induced stimulus change on human discrimination. *Journal of Experimental Psychology,* 1967, **74,** 453–456.

Treichler, F. R., & Way, S. J. Task variables and the effects of response-contingent stimulus change on discrimination performance. *Journal of Experimental Psychology,* 1968, **76,** 671–673.

Trzeworski, A., & Teune, H. *The logic of comparative social inquiry.* New York: Wiley-Interscience, 1970.

Tulving, E. Theoretical issues in free recall. In T. R. Dixon and D. L. Horton (Eds.), *Verbal behavior and general behavior theory.* Englewood Cliffs, N. J.: Prentice-Hall, 1968.

Tulving, E. Episodic and semantic memory. In E. Tulving & W. Donaldson (Eds.), *Organization of memory.* New York: Academic Press, 1972.

Tulving, E., & Bower, G. H. The logic of memory representations. In G. H. Bower (Eds.), *The psychology of learning and motivation: Advances in research and theory.* Vol. 8. New York: Academic Press, 1974.

Tulving, E., & Donaldson, W. (Eds.), *Organization of memory.* New York: Academic Press, 1972.

Tulving, E., & Psotka, J. Retroactive inhibition and free recall: Inaccessibility of information available in the memory store. *Journal of Experimental Psychology,* 1971, **87,** 1–8.

Tulving, E., & Thomson, B. M. Encoding specificity and retrieval processes in episodic memory. *Psychological Review,* 1973, **80,** 352–373.

Underwood, B. J. Interference and forgetting. *Psychological Review,* 1957, **64,** 49–60.

Underwood, B. J. Attributes of memory. *Psychological Review,* 1969, **76,** 559–573.

Underwood, B. J., & Postman, L. Extraexperimental sources of interference in forgetting. *Psychological Review,* 1960, **67,** 73–95.

Underwood, B. J., & Schulz, R. W. *Meaningfulness and verbal learning.* Chicago: Lippincott, 1960.

Vaughan, J., & Schrier, A. M. Effect of locus of reinforcement on learning set formation by rhesus monkeys. *Learning & Motivation,* 1970, **1,** 79–85.

Wagner, A. R. Elementary associations. In H. H. Kendler & J. T. Spence (Eds.), *Essays in neobehaviorism: A memorial volume to Kenneth W. Spence.* New York: Appleton-Century-Crofts, 1971.

Wagner, A. R. Priming in STM: An information-processing mechanism for self-generated or retrieval-generated depression in performance. In T. J. Tighe & R. N. Leaton (Eds.), *Habituation: Perspectives from child development, animal behavior, and neurophysiology.* Hillsdale, N. J.: Laurence Erlbaum Assoc., 1976.

Wagner, A. R., Rudy, J. W., & Whitlow, J. W. Rehearsal in animal conditioning. *Journal of Experimental Psychology,* 1973, **97,** 407–426.

Wagner, A. R., & Terry, W. S. Backward conditioning to a CS following an expected vs. a surprising UCS. *Animal Learning & Behavior,* 1976, in press.

Warner, L. H. The association span of the white rat. *Journal of Genetic Psychology,* 1932, **41,** 57–90.

Watkins, M., & Tulving, E. Episodic memory: When recognition fails. *Journal of Experimental Psychology: General,* 1975, **104,** 5–29.

Waugh, N. C. Presentation time and free recall. *Journal of Experimental Psychology,* 1967, **73,** 39–44.

Weiner, B. *Theories of motivation.* Chicago: Markham, 1972.

Weisinger, R. S., Parker, L. F., & Skorupski, J. D. Conditioned taste aversions and specific need states in the rat. *Journal of Comparative & Physiological Psychology,* 1974, 87, 655–660.

Weiskrantz, L. (Ed.) *Analysis of behavioral change.* New York: Harper & Row, 1968.

Weiskrantz, L. Memory. In L. Weiskrantz (Ed.), *Analysis of behavioral change.* New York: Harper & Row, 1968.

Wickelgren, W. A. Single-trace fragility theory of memory dynamics. *Memory & Cognition,* 1974, **2,** 775–780.

Wickens, D. D. Encoding categories of words: An empirical approach to meaning. *Psychological Review,* 1970, **77,** 1–15.

Wickens, D. D., Hall, J., & Reid, L. S. Associative and retroactive inhibition as a function of the drive stimulus. *Journal of Comparative & Physiological Psychology,* 1949, **42,** 398–403.

Wilcoxon, H. C., Dragoin, W. B., & Kral, P. A. Illness-induced aversions in rats and quail: Relative salience of visual and gustatory cues. *Science,* 1971, **171,** 826–828.

Williams, M. Traumatic retrograde amnesia and normal forgetting. In C. A. Talland & N. C. Waugh (Eds.), *The pathology of memory.* New York: Academic Press, 1969. Pp. 75–80.

Wilson, E. H. Reinstatement of extinguished avoidance in the rat: Feeling states as US-specific mediators. Unpublished doctoral dissertation, Department of Psychology, Indiana University, Bloomington, 1971.

Winograd, E. Some issues relating animal memory to human memory. In W. K. Honig & P. H. R. James (Eds.), *Animal memory.* New York: Academic Press, 1971.

Wood, G. Organizational processes and free recall. In E. Tulving & W. Donaldson (Eds.), *Organization of memory.* New York: Academic Press, 1972.

Woods, R. T. & Piercy, M. A similarity between amnesic memory and normal forgetting. *Neuropsychologia,* 1974, **12,** 437–445.

Woodbury, C. B. The learning of stimulus patterns by dogs. *Journal of Comparative Psychology,* 1943, **35,** 29–40.

Woodson, W. E., & Conover, D. W. *Human engineering guide for equipment designers.* (2nd. rev. ed.) Berkeley: University of California Press, 1965.

Worsham, R. W. Delayed matching-to-sample as temporal discrimination in monkeys. Unpublished doctoral dissertation, Rutgers University, New Brunswick, N.J., 1973.

Worsham, R. W. Temporal discrimination factors in the delayed matching-to-sample task in monkeys. *Animal Learning & Behavior,* 1975, **3,** 93–97.

Worsham, R. W., & D'Amato, M. R. Short-term memory in monkeys: Storage vs. discrimination. Paper presented at the Psychonomic Society, St. Louis, Missouri, 1973.

Wyrwicka, W. Studies of motor conditioned reflexes. VI. On the effect of experimental situation upon the course of motor conditioned reflexes. *Acta Biologiae Experimentalis.,* 1956, **17,** 189–203.

Yerkes, R. M., & Yerkes, D. N. Concerning memory in the chimpanzee. *Journal of Comparative Psychology,* 1928, **8,** 237–271.

Zentall, T. R. Effect of context change in forgetting in rats. *Journal of Experimental Psychology,* 1970, **86,** 440–448.

Zentall, T. R. Memory in the pigeon: Retroactive inhibition in a delayed matching task. *Bulletin of the Psychonomic Society,* 1973, **1,** 126–128.

Zentall, T. R., & Hogan, D. E. Memory in the pigeon: Proactive inhibition in a delayed matching task. *Bulletin of the Psychonomic Society,* 1974, **4,** 109–112.

Indices

Author Index

Subject Index

N

O

P